◎ 陈 涛 主 编　孔繁玉 孙鹏军 张炉焱 副主编

虚拟化与容器技术

清华大学出版社
北京

内容简介

本书通过深入浅出的方式介绍KVM虚拟化技术与Docker容器技术的概念、原理及实现方法，内容包括KVM概述、安装KVM、创建KVM虚拟机、虚拟机管理、管理KVM虚拟网络、管理KVM虚拟存储、容器技术简介、Docker镜像管理、Docker容器管理、Docker 网络管理、Docker存储管理、使用Dockerfile创建镜像等，并包含丰富的实验和案例，内容丰富，结构清晰，案例典型，实践性强。

本书既可作为希望了解虚拟化技术和容器技术的系统管理员、DevOps工程师的技术入门书籍，也可作为高等院校云计算相关课程的教材或教学参考书。

图书在版编目(CIP)数据

虚拟化与容器技术 / 陈涛主编. 一北京：清华大学出版社，2023.9
ISBN 978-7-302-64400-2

I. ①虚… II. ①陈… III. ①Linux操作系统—虚拟处理机 IV. ①TP338

中国国家版本馆CIP数据核字(2023)第149539号

责任编辑：刘金喜
封面设计：高娟妮
版式设计：孔祥峰
责任校对：成凤进
责任印制：刘海龙

出版发行：清华大学出版社
网　　址：http://www.tup.com.cn，http://www.wqbook.com
地　　址：北京清华大学学研大厦A座　　**邮　　编**：100084
社 总 机：010-83470000　　**邮　　购**：010-62786544
投稿与读者服务：010-62776969，c-service@tup.tsinghua.edu.cn
质 量 反 馈：010-62772015，zhiliang@tup.tsinghua.edu.cn
印 装 者：北京同文印刷有限责任公司
经　　销：全国新华书店
开　　本：185mm×260mm　　**印　　张**：17.5　　**字　　数**：448千字
版　　次：2023年9月第1版　　**印　　次**：2023年9月第1次印刷
定　　价：69.80元

产品编号：100099-01

PREFACE

近年来，云计算技术在全球范围内迅猛发展，已经成为数字化时代的重要基础设施，深入到人们的生活和工作中。数字化、网络化、智能化、协同化已经成为世界经济和科技发展的主要趋势，我国正在逐步加强数字基础设施建设，积极发展新型基础设施，以进一步推进云计算、大数据、人工智能等新一代信息技术产业的发展。可以说，在当前数字化转型的背景下，云计算已经成为推动各行各业创新发展的重要动力。

KVM 虚拟化技术与 Docker 容器技术是云计算的两个重要“引擎”，它们分别在不同领域中具有独特的优势与适用场景。KVM 虚拟化是一种基于内核的虚拟机技术，在系统虚拟化方面有着广泛的应用。Docker 容器则是一个颠覆性的容器化技术，具有快速迁移、可移植性强等诸多优点。本书将围绕 KVM 虚拟化与 Docker 容器展开，探究它们的基本原理、适用场景、实现与管理等方面的知识，帮助读者更全面、更深入地了解这两项技术的特点以及如何使用它们来助力业务发展。

本书内容

本书共分 13 章，分为两部分：第一部分是 KVM 虚拟化技术(第 1～第 6 章)，第二部分是 Docker 容器技术(第 7～第 13 章)。

第 1 章介绍虚拟化的定义与历史，KVM 的原理、功能与优势。

第 2 章讲解 KVM 的架构，如何构建实验环境，如何在 Linux 宿主机中安装部署 KVM。

第 3 章讲解如何通过图形化管理工具 virt-manager 和命令行管理工具 virt-install 创建虚拟机。

第 4 章讲解如何通过 virsh 命令管理虚拟机，包括创建、暂停、恢复、停止及删除等生命周期管理的操作。

第 5 章讲解虚拟网络的管理，包括 NAT、桥接网络的原理与配置。

第 6 章讲解虚拟存储的管理，包括存储池、存储卷的原理与日常管理。

第 7 章介绍容器的定义、部署、发展历史及应用场景等。

第 8 章讲解 Docker 的架构原理、常用子命令、镜像及镜像仓库的使用。

第 9 章讲解 Docker 容器的基本概念及常用操作命令。

第 10 章讲解 Docker 网络原理，容器与外部网络的相互访问原理，默认网络类型分析，常用管理命令等。

第 11 章讲解 Docker 存储概述、存储卷的管理及适用场景。

第 12 章讲解 Dockerfile 的常用配置指令，如何使用 Dockerfile 创建镜像。

第 13 章讲解 Docker 实战案例：Linux 操作系统镜像、Nginx/Apache HTTPD 服务、MySQL、MariaDB、MongoDB 数据库服务。

适用对象

本书既可作为希望了解虚拟化技术和容器技术的系统管理员、DevOps 工程师的技术入门书籍，也可作为高等院校云计算相关课程的教材或教学参考书。在学习本书之前，应先具备一定的 Linux 基础知识。在学习过程中，既要充分理解相关的概念和原理，也要注重实践和动手操作，以加深对知识的理解和掌握。掌握 KVM 虚拟化技术与 Docker 容器技术，能够为后续学习 OpenStack、K8S 等云平台打下坚实基础。本书所有的实验及案例，均在 CentOS Stream 9 Linux 环境中经过验证。本书作为教材使用时，建议课时安排 64～72 学时。

本书由陈涛任主编，孔繁玉、孙鹏军、张炉焱任副主编，编写团队既有多年项目研发的历练，又有一线教学经验。本书编写过程中参考了国内外一些相关书籍，谨向这些作者表示诚挚的感谢。由于时间仓促，加之编者水平有限，书中难免存在不足之处，敬请广大读者批评指正。

本书 PPT 教学课件和案例源文件可通过扫描下方二维码下载。

服务邮箱：476371891@qq.com。

教学资源

编者

2023 年 3 月

CONTENTS

第 1 章

KVM概述

现今人们已经越来越离不开互联网。在互联网上，云计算技术已经非常成熟。虚拟化是云计算的核心技术之一，也是云计算的底层支撑技术之一。利用虚拟化技术，可以实现 IT 资源弹性分配，使 IT 资源分配更加灵活，满足多样化的应用需求。现如今，KVM(kernel-based virtual machine，基于内核的虚拟机)可以在大多数 Linux 发行版上运行，并且 KVM 也是很多公共云服务商的默认管理程序。

本章要点

◎ 虚拟化概述

◎ KVM 概述

◎ KVM 的优点

1.1 虚拟化技术简介

要学好 KVM，就需要掌握一些与虚拟化相关的基础知识，包括虚拟化的概念、发展史和实现方式。

1.1.1 虚拟化的定义

虚拟化是一个广义的术语，通常是指将计算机原件在虚拟的物理机上运行而不是在真实的物理机上运行。通俗来讲，就是可以将单 CPU 模拟成多个 CPU 并行，允许一个平台同时运行多个操作系统，并且应用程序都可以在互相独立的空间内运行而互不影响。

1.1.2 操作系统虚拟化的发展历史

操作系统虚拟化通常表现为在单一系统上运行多个虚拟操作系统，这些虚拟操作系统同时运行，而且又相互独立。

操作系统虚拟化技术的发展史如图 1-1 所示。从图 1-1 中可看出，20 世纪 60 年代就已经出现了虚拟化技术，最早是在 IBM CP-40 大型机上使用了虚拟内存和虚拟机的概念。

除此之外，IBM 公司还有很多与虚拟化有关的创新贡献。

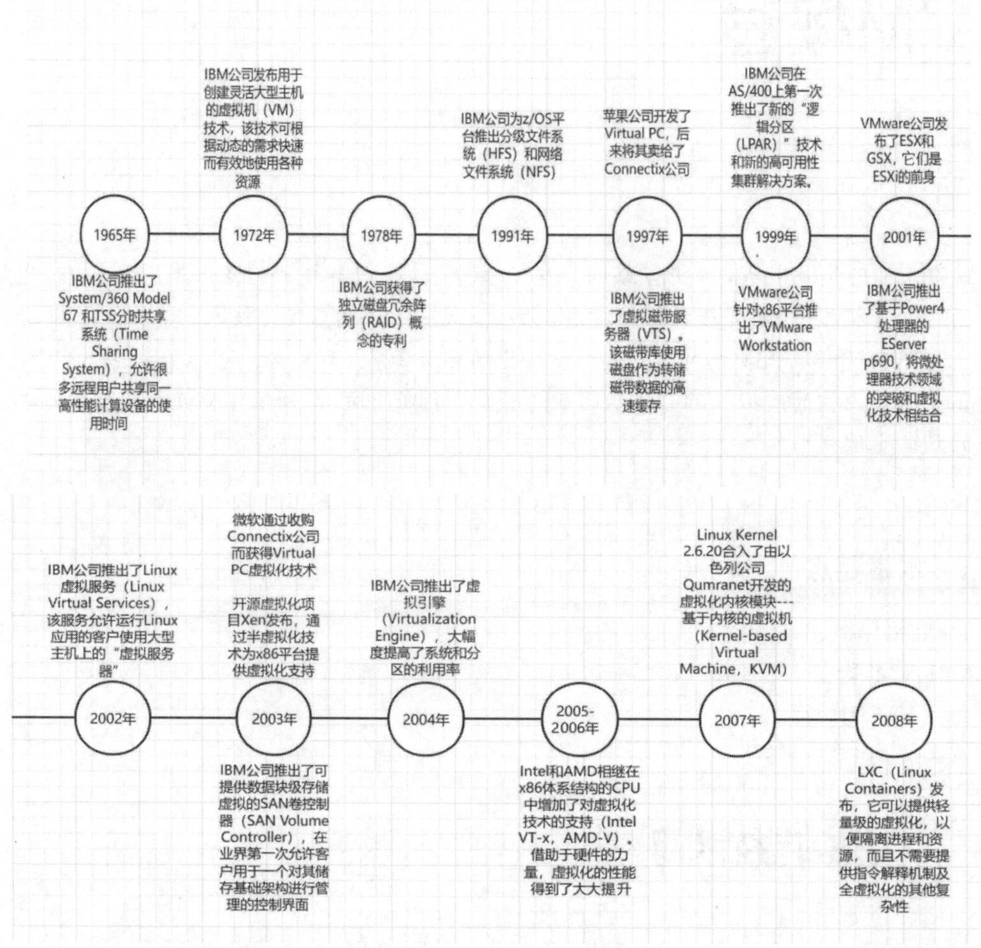

图 1-1

1.1.3 虚拟化的分类

虚拟化的方式多种多样，耳熟能详的名字有：全虚拟化、类虚拟化、硬件虚拟化、混合虚拟化等。这些不同的虚拟化方式，并不是根据同一个标准来分类的。

1. 根据虚拟化平台划分

从虚拟化平台的角度来划分，可以分为全虚拟化和类虚拟化。

(1) 全虚拟化：指虚拟机监视程序(virtual machine monitor，VMM)虚拟出来的平台是显示存在的平台，客户机并不知道自己运行在虚拟平台上，全虚拟化中的客户机操作系统是不需要做任何修改的。

(2) 类虚拟化：指通过对客户机进行源代码级的修改，让客户机可以使用虚拟化的资源。类虚拟化一般也被用来优化 I/O，客户机的操作系统通过高度优化的 I/O 协议，可以和 VMM 紧密结合达到近似于物理机的速度。

2. 根据应用对象划分

根据应用对象的不同，虚拟化技术可以分为以下几种。

(1) 操作系统虚拟化：将资源划分为一个或多个执行环境，以运行不同的操作系统。

(2) 存储虚拟化：对存储资源进行抽象化表现，从早期的 RAID、LVM 到后来的 VTL(虚拟化磁带库)，以及现在软件定义的存储等都是存储虚拟化。

(3) 网络虚拟化：软件定义的网络(software defined network，SDN)主要分为两个方向：控制平面虚拟化与数据平面虚拟化。

(4) GPU 虚拟化：将图形处理器单元(graphics processing unit，GPU)虚拟化之后，可以让运行在数据中心服务器上的虚拟机实例共享使用同一块或多块 GPU 处理器。

(5) 软件虚拟化：与常用的“绿色”软件类似，可以大大降低企业中软件部署、维护、升级的综合成本

(6) 硬件虚拟化：例如 SR-IOV(single root I/O virtualization)技术对网卡等设备进行封装、管理、共享，在其上创建多个 VF(virtual function)，每个 VF 对操作系统来讲都是真实的物理设备。

1.1.4 虚拟化的目的

虚拟化的主要目的是对 IT 基础设置进行简化，以及对资源进行访问。上面我们已经讲解过了虚拟化的原理，所以这里不再赘述。

虚拟化使用软件的方法重新定义及划分 IT 资源，可以实现 IT 资源的动态分配、灵活调度、跨域共享，提高 IT 资源利用率，使 IT 资源能够真正成为社会基础设施，服务于各行各业。

与传统 IT 资源分配的应用方式相比，虚拟化具有以下优势：

(1) 虚拟化技术可以大大提高资源的利用率，提供相互隔离、安全、高效的应用环境。

(2) 虚拟化系统能够方便地管理和升级资源。

虚拟化技术等的发展促进了云计算技术的飞速发展，也可以说虚拟化是云计算的基础，没有虚拟化就没有云计算。

1.2 KVM 简介

KVM 既可以指 KVM 技术本身，也可以指 KVM 开源虚拟化解决方案。与 VMware、Microsoft 公司的商业虚拟化产品不同，KVM 虚拟化解决方案是由 KVM、QEMU 和 libvirt 这 3 个既独立又合作的开源项目组合在一起形成的。

1.2.1 什么是 KVM

KVM(用于基于内核的虚拟机)是用于 x86 硬件上的 Linux 的完整虚拟化解决方案，包含虚拟化扩展(Intel VT 或 AMD-V)。它由一个可加载的内核模块 kvm.ko 和一个处理器特定模块 kvm-intel.ko 或 kvm-amd.ko 组成，该模块提供核心虚拟化基础设施。

使用 KVM，可以运行多个未修改的 Linux 或 Windows 镜像的虚拟机。每个虚拟机都有私有的虚拟化硬件：网卡、磁盘、图形适配器等。

KVM 是开源软件。从 KVM 2.6.20 开始，KVM 的内核组件包含在主线 Linux 中。 从 KVM 1.3 开始，KVM 的用户空间组件包含在主线 QEMU 中。

由此可以看出，KVM 是基于内核的虚拟机。

1.2.2 KVM 的发展史

根据维基百科的记载：

(1) 阿维・齐维迪(Avi Kivity)在一家名为 Qumranet 的初创企业开始 KVM 的研发工作，随后此公司于 2008 年被 Red Hat(红帽)公司收购。

(2) KVM 被合并入 Linux 内核版本 2.6.20 的主流分支，于 2007 年 2 月 5 日发布。

(3) KVM 现由保罗・邦齐尼(Paolo Bonzini)维护。

1.2.3 KVM 历史大事件

2005 年 11 月，Intel 发布正式支持 VT 技术的处理器。

2006 年 5 月，AMD 发布支持 AMD-V 技术的处理器。

KVM 虚拟机最初由一个以色列的创业公司 Qumranet 开发，作为它们的 VDI 产品的虚拟机。为了简化开发，KVM 的开发人员并没有选择从底层开始新写一个 Hypervisor，而是选择了基于 Linux kernel，通过加载新的模块从而使 Linux Kernel 本身变成一个 Hypervisor。

2006 年 8 月，在先后完成了基本功能、动态迁移以及主要的性能优化之后，Qumranet 正式对外宣布 KVM 的诞生并推向 Linux 内核社区。同年 10 月，KVM 模块的源代码被正式接纳进入 Linux Kernel(内核)，成为内核源代码的一部分。

2007 年 2 月发布的 Linux 2.6.20 是第一个带有 KVM 模块的 Linux 内核正式发布版本。

2008 年 9 月 4 日，同内核社区保持着很深渊源的著名 Linux 发行版提供商——Redhat 公司出人意料地收购 Qumranet，从而成为 KVM 开源项目的新东家，投入较多资源在 KVM 虚拟化开发中。

2010 年 11 月，Redhat 公司推出了新的企业版 Linux——RHEL 6，在这个发行版本中集成了最新的 KVM 虚拟机，而去掉了在 RHEL 5.x 系列中集成的 Xen。KVM 成为 RHEL 默认的虚拟化方案。

1.2.4 KVM 的功能

KVM 所支持的功能包括：

- 支持 CPU 和内存超额分配(overcommit)
- 支持半虚拟化 I/O(virtio)

- 支持热插拔(CPU、块设备、网络设备等)
- 支持对称多处理(symmetric multi-processing，SMP)
- 支持实时迁移(live migration)
- 支持 PCI 设备直接分配和单根 I/O 虚拟化(SR-IOV)
- 支持内核同页合并(kernel samepage merging，KSM)
- 支持 NUMA (non-uniform memory access，非一致存储访问结构)

1.2.5 KVM 的优势

KVM 是 Linux 的一部分，Linux 也是 KVM 的一部分。Linux 有的优势，KVM 全都有。然而，KVM 的某些特点让它成为了企业的首选虚拟机监控程序。

1. 安全性

KVM 利用安全增强型 Linux(SELinux)和安全虚拟化(sVirt)组合来加强虚拟机的安全性和隔离性。SELinux 在虚拟机周围建立安全边界。sVirt 则扩展 SELinux 的功能，使强制访问控制(MAC)安全机制应用到客户虚拟机，并预防手动标记错误。

2. 存储方便

KVM 能够使用 Linux 支持的任何存储，包括某些本地磁盘和网络附加存储(NAS)。还可以利用多路径 I/O 来增强存储并提供冗余能力。KVM 还支持共享文件系统，因此虚拟机镜像可以由多个主机共享。磁盘镜像支持精简置备，可以按需分配存储，不必预先备妥一切。

3. 硬件支持

KVM 可以使用多种多样的认证 Linux 兼容硬件平台。由于硬件供应商经常助力内核开发，所以 Linux 内核中通常能快速采用最新的硬件功能。

4. 内存管理

KVM 继承了 Linux 的内存管理功能，包括非统一内存访问和内核同页合并。虚拟机的内存可以交换，也可通过大型宗卷支持来提高性能，还可由磁盘文件共享或支持。

5. 实时迁移

KVM 支持实时迁移，也就是能够在物理主机之间移动运行中的虚拟机，而不会造成服务中断。虚拟机保持开机状态，网络连接保持活跃，各个应用也会在虚拟机重新定位期间正常运行。KVM 也会保存虚拟机的当前状态，从而存储下来供日后恢复。

6. 性能和可扩展性

KVM 继承了 Linux 的性能，针对客户机和请求数量的增长进行扩展，满足负载的需求。KVM 可让要求最苛刻的应用工作负载实现虚拟化，而这也是许多企业虚拟化设置的基础，如数据中心和私有云(基于 OpenStack)等。

7. 调度和资源控制

在 KVM 模型中，虚拟机是一种 Linux 进程，由内核进行调度和管理。通过 Linux 调度程序，可对分配给 Linux 进程的资源进行精细控制，并且保障特定进程的服务质量。在 KVM 中，这包括完全公平的调度程序、控制组、网络命名空间和实时扩展。

8. 更低延迟，更高优先级

Linux 内核提供实时扩展，允许基于虚拟机的应用以更低的延迟、更高的优先级来运行(相对于裸机恢复)。内核也将需要长时间计算的进程划分为更小的组件，再进行相应的调度和处理。

1.3 习题

1. 用自己的话描述什么是虚拟化。
2. 用自己的话描述什么是 KVM。
3. KVM 有哪些特点？

第 2 章

安装KVM

第 1 章介绍了虚拟化的基本概念和 KVM 的基础知识。本章将讲解如何在 Windows 系统中安装虚拟机。

本章要点

- ◎ VMware Workstation 16 Pro 的安装
- ◎ 图形化下安装 KVM 虚拟机
- ◎ 最小化下安装 KVM 虚拟机

2.1 KVM 架构解析

KVM 是在硬件辅助虚拟化技术之上构建起来的虚拟机监视器，KVM 对硬件的最低依赖是 CPU 的硬件虚拟化支持，而其他的内存和 I/O 的硬件虚拟化支持会让整个 KVM 虚拟化下的性能得到更多的提升。

KVM 虚拟化的核心主要由 KVM 内核模块和 QEMU 用户态工具两个模块组成，下面将分别对这两个模块进行讲解。

2.1.1 KVM 内核模块

KVM 内核模块属于标准 Linux 内核的一部分，KVM 内核是一个专门提供虚拟化功能的模块，主要负责 CPU 和内存的虚拟化，其中包括了客户机的创建、虚拟内存的分配、CPU 执行模式的切换、vCPU 寄存器的访问、vCPU 的执行。

KVM 模块是 KVM 虚拟化的核心模块，它在 Linux 内核中由两部分组成，分别是处理器架构无关的部分和处理器架构相关的部分，其中无关的部分可以通过 lsmod 命令进行查看，这一部分也叫作 KVM 模块。在 Intel 平台上的 kvm_intel 内核模块就是与处理器架构相关的部分。

KVM 的主要功能是初始化 CPU 硬件，打开虚拟化模式，然后将虚拟客户机运行在虚拟机上，并对虚拟客户机提供一定的支持。

2.1.2 QEMU 用户态工具

QEMU 是纯软件设计的虚拟化模拟器，QEMU 几乎可以模拟任何硬件设备，我们最熟悉的功能就是模拟一台能独立运行操作系统的虚拟机。

QEMU 是纯软件实现的，所以它所有的指令都要经过 QEMU，从而也导致了性能降低。我们在上面讲到 KVM 是硬件辅助的虚拟化技术，所以在生产环境中，QEMU 一般都是配合着 KVM 来完成虚拟化工作的，由 KVM 完成繁琐的 CPU 虚拟化和内存虚拟化，由 QEMU 负责 I/O 设备的虚拟化，这样能有效地发挥它们的优势。

从本质上来看，虚拟出来的每个虚拟机对应宿主机上的 QEMU 进程，而虚拟机的执行线路即 CPU 线路和 I/O 线路对应 QEMU 进程中的一个线程。

2.2 安装环境准备

“工欲善其事，必先利其器。”要学习 KVM，就一定要先把环境准备好，有了学习的环境，才能跟着步骤或者案例一步一个脚印地往前走。

2.2.1 Windows 环境准备

在安装 KVM 之前，需要先对宿主机进行一些设置，这些设置必须保持开启状态，否则将无法进行虚拟化的实验。实验环境如下：

- 物理机操作系统：Windows 10(专业版)。
- 虚拟化软件：VMware Workstation 16 Pro。
- 虚拟机操作系统：CentOS Stream 8 Linux 发行版本。

接下来首先讲解在物理机上需要修改的一些设置。

如果物理机需要重新安装操作系统，推荐使用 MSDN 官网下载系统镜像。

在物理机中需要进入到 BIOS 中将 VT(virtualization technology，虚拟化技术)的选项设置为 Enable。根据电脑品牌的不同，进入 BIOS 的方法也不同，但都需要在物理机刚按下开机键的情况下连续按对应的键才能进入。台式机根据主板的类型不同，进入 BIOS 所按的键也不同，具体各型号进入 BIOS 的按键如表 2-1 所示。

表 2-1　不同的台式机进入 BIOS 设置对应的按键

BIOS 型号	进入 CMOS SETUP 的按键	屏幕是否有提示
AMI	DEL 键或者 ESC 键	有
AWARD	DEL 键或者 Ctrl+Alt+ESC 键	有

(续表)

BIOS 型号	进入 CMOS SETUP 的按键	屏幕是否有提示
MR	DEL 键或者 Ctrl+Alt+ESC 键	无
Quadtel	F2 键	有
HP	F2 键	有
Phoenix	Ctrl+Alt+S 键	无
AST	Ctrl+Alt+ESC 键	无
COMPAQ	屏幕右上角出现光标时按 F10 键	无

不同品牌的笔记本进入 BIOS 也有所不同，具体按键如表 2-2 所示。

表 2-2 不同品牌笔记本进入 BIOS 对应的按键

笔记本品牌	进入 CMOS SETUP 的按键
ThinkPad/IBM	冷启动按 F1，部分新型号可以在重启时按 F1 键
SONY	启动或重启时按 F2 键
Dell	启动或重启时按 F2 键
HP	启动或重启时按 F2 键
Acer	启动或重启时按 F2 键
Toshiba	冷启动时按 ESC 键然后按 F1 键
大多数中国大陆生产和中国台湾地区的品牌	启动或重启时按 F2 键

注：进入 BIOS 的按键不同是由主板厂商及主板决定的。

进入 BIOS 后，通过观察判断 BIOS 的版本，如果是 Phoenix BIOS(图 2-1)，那么只需要找到“Configuration”，然后找到“Intel Virtual Technology”选项，将此选项设置为“Enable”即可，其中 Enable 为启用，Disable 为禁用。

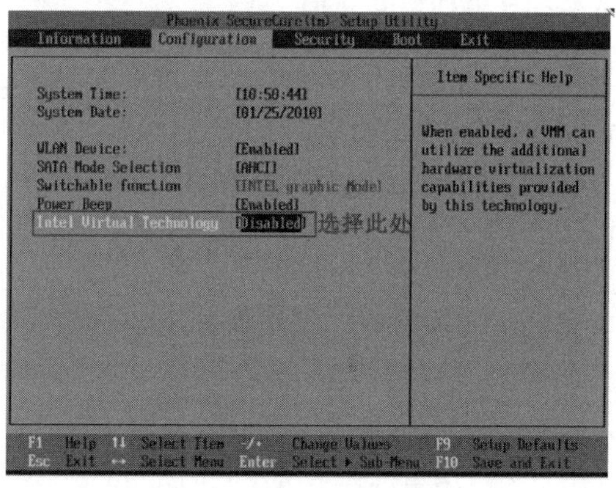

图 2-1 Phoenix BIOS 版本

如果 BIOS 的版本为 Award BIOS(图 2-2)，则需要找到“Advanced BIOS Features”选项，按键盘上的 Enter 键，进入到如图 2-3 所示的页面，找到“Virtualization Technology”选项，将

此选项的值设置为“Enable”，然后按键盘上的“F10”键保存当前的更改，重启计算机即可开启物理机的 VT(虚拟化技术)。

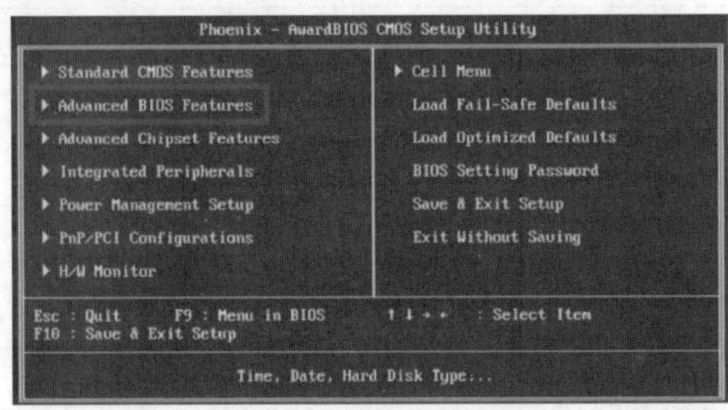

图 2-2　Award BIOS 中的 Advanced BIOS Features 选项

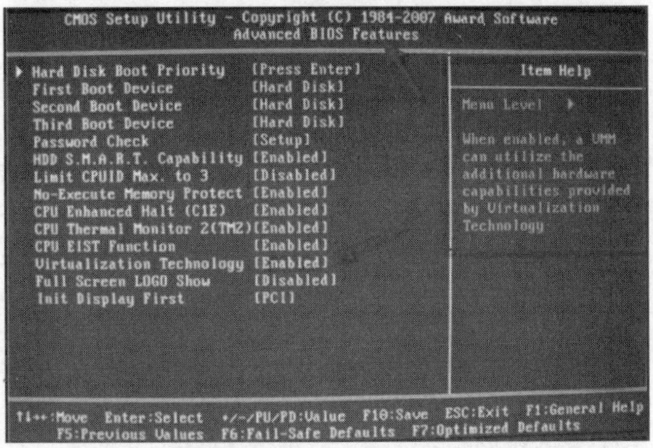

图 2-3　Award BIOS 开启 VT

除上述两种品牌之外的其他品牌，请自行查找“电脑主机品牌名或主板名如何开启 VT”，如果是联想电脑则可以直接查询“联想笔记本怎么打开 VT”，这里就不再过多地赘述如何开启 VT。

2.2.2　下载与安装 VMware Workstation

实验环境使用的是 VMware Workstation 16 Pro，大家可以根据自己的需求选择需要的版本，这里因为物理机是 Windows 10 操作系统，所以直接单击 Workstation 16 Pro for Windows 下面的“立即下载”进行下载。

VMware Workstation 官网下载界面如图 2-4 所示，单击图片左下角的“立即下载”按钮进行 VMware Workstation 安装程序的下载，下载完成后可以看到文件 VMware-workstation-full-16.2.4-20089737.exe。接下来以这个安装包为例，进行详细的安装过程的讲解。

在页面的最底端有两个“立即下载”按钮，单击左边的“立即下载”将下载 Windows 版本的 VMware Workstation，单击右侧的“立即下载”所下载的是 Linux 版本的 VMware Workstation，因为是要在 Windows 上使用，所以这里选择单击左侧的“立即下载”。

下载结束后就可以双击安装包开始安装了，图 2-5 所示的页面就是打开安装包之后的页面，下面就开始进行 VMware Workstation 的安装。

图 2-4 Vmware Workstation 官网下载

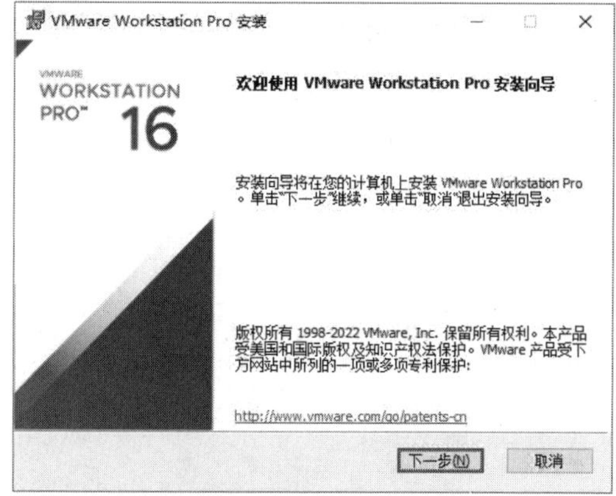

图 2-5 Vmware Workstation 安装向导

如图 2-6 所示为 VMware Workstation 的用户许可协议，任何软件都有用户许可协议，这里需要勾选“我接受许可协议中的条款”，之后单击“下一步”，如果需要详细了解该许可协议，可以打印出来仔细阅读。

注意

在如图 2-7 所示的窗口中，可以修改安装位置，一般来说，除了系统驱动等系统必要的软件会安装在系统的 C 盘内，其他的软件一般都会选择安装在非 C 盘的其他盘中，这样不会影响系统的启动和运行。单击其中的“更改”按钮，可以选择一个非 C 盘的其他位置。因笔者电脑上只有 C 盘，所以这里不做修改。

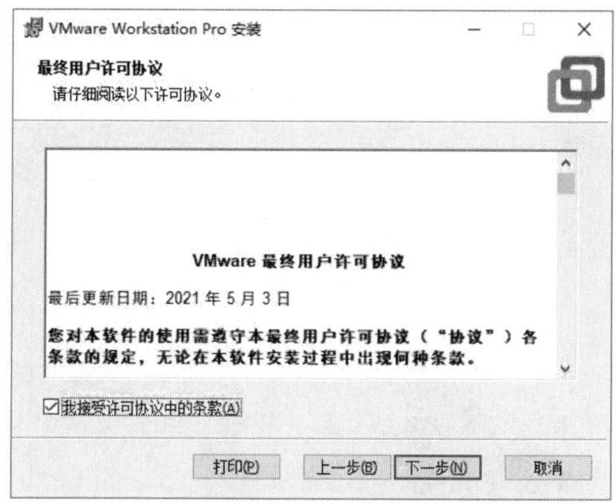

图 2-6 许可协议

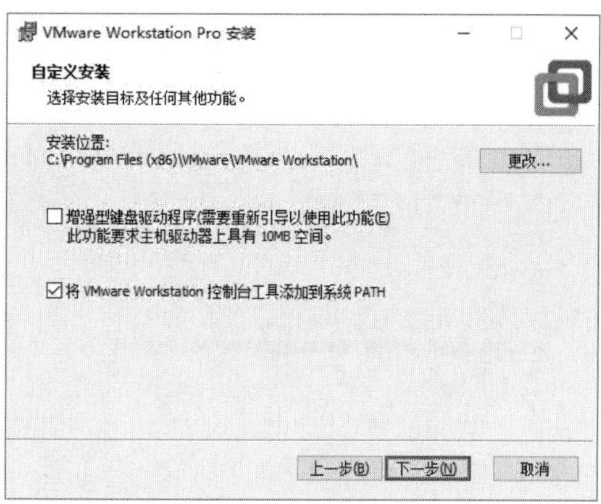

图 2-7 选择安装目标及其他功能

更改完安装位置后，单击“下一步”继续安装。

这里可以根据自己的需求选择是否要及时更新产品，如果不想及时更新产品，可以选择不勾选“启动时检查产品更新”，如图 2-8 所示。

在接下来打开的图 2-9 所示的窗口中，根据自己的需求进行选择，如果选择默认设置，会在我们的电脑桌面上和“开始”菜单栏里面加入 VMware Workstation 的快捷启动方式。

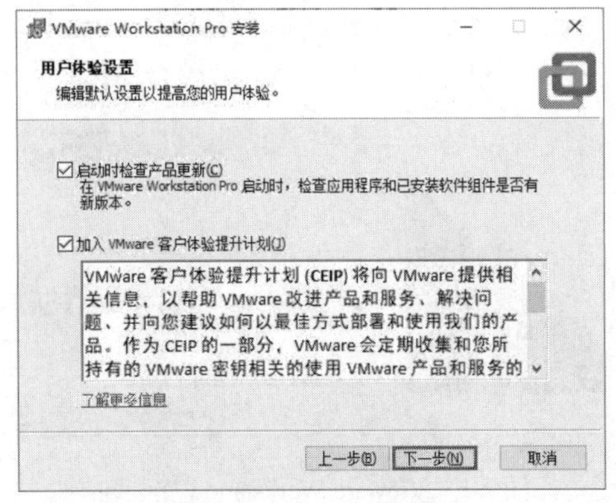

图 2-8 用户体验设置

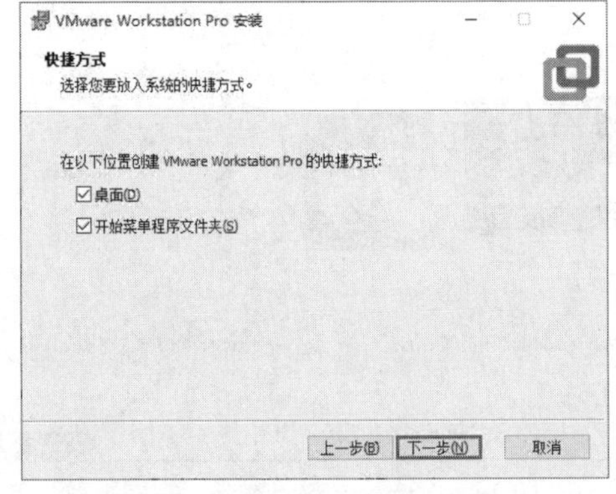

图 2-9 快捷方式

到这一步所有的准备工作就都已经做好了，这里只需要单击“安装”就行(图 2-10)，然后耐心等待程序的安装(图 2-11)，直到出现如图 2-12 所示的窗口，就表示安装完成了。

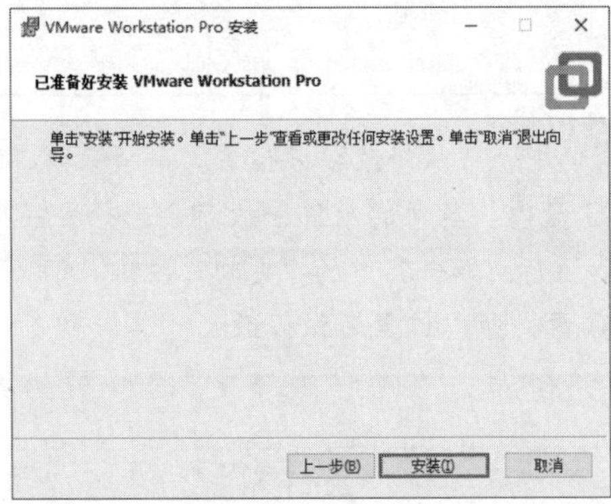

图 2-10 准备安装

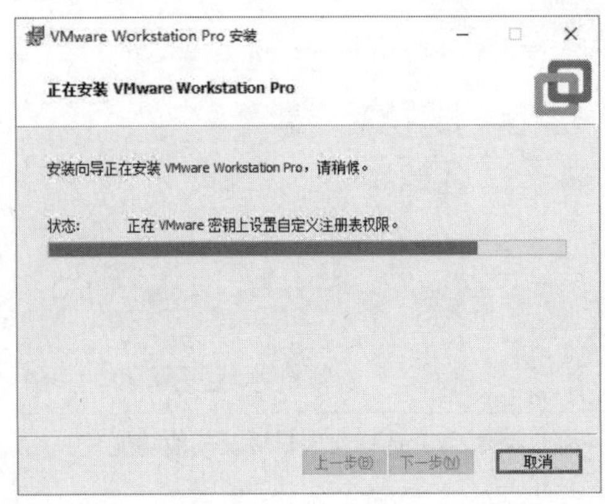

图 2-11 正在安装

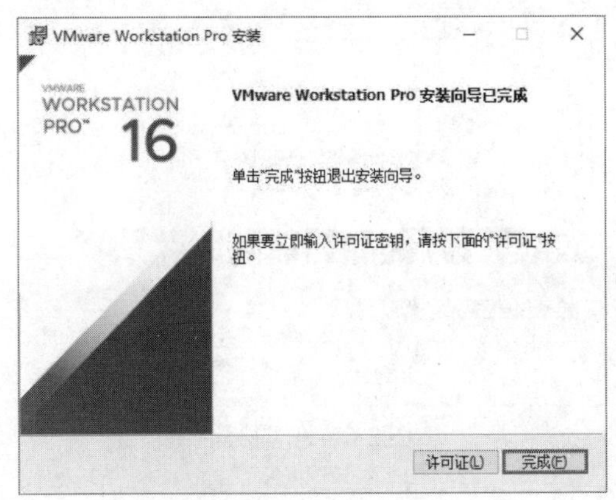

图 2-12 安装完成

安装完成后，查看计算机网络属性的变化，这对以后理解虚拟网络会有帮助。系统新增两个网络连接，分别是 Vmware Network Adapter Vmnet1 和 Vmware Network Adapter Vmnet8，如图 2-13 所示。它们分别与 Vmware Workstation 软件中的仅主机(Host-Only)模式和 NAT 模式的虚拟机相连接。

再查看系统服务，系统新增 4 个与 Vmware Workstation 软件相关的服务，如图 2-14 所示。

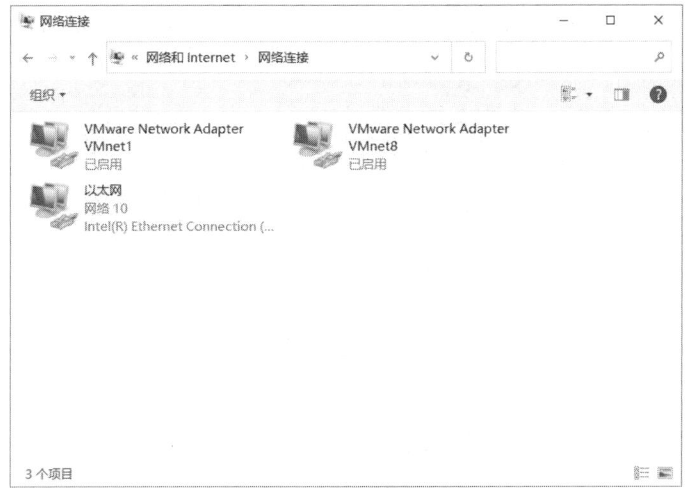

图 2-13　新增加的网络连接

服务

文件(F)　操作(A)　查看(V)　帮助(H)

服务(本地)

VMware Authorization Service

停止此服务
重启动此服务

描述:
用于启动和访问虚拟机的授权及身份验证服务。

名称	描述	状态	启动类型	登录为
Virtual Disk	提供用于磁盘、卷、文件系统和存储阵列的管理服务。		手动	本地系...
VMware Authorization Service	用于启动和访问虚拟机的授权及身份验证服务。	正在...	自动	本地系...
VMware DHCP Service	DHCP service for virtual networks.	正在...	自动	本地系...
VMware NAT Service	Network address translation for virtual networks.	正在...	自动	本地系...
VMware USB Arbitration Service	Arbitration and enumeration of USB devices for virtual machines	正在...	自动	本地系...
Volume Shadow Copy	管理并执行用于备份和其它用途的卷影复制。如果此服务被终止，备份...		手动	本地系...
W3C 日志记录服务	为 Internet Information Services (IIS)提供 W3C 日志记录。如果此...		手动	本地系...
Wacom Professional Service	Driver for Wacom Tablets	正在...	自动	本地系...
WalletService	电子钱包客户端使用的主机对象		手动	本地系...
Wallpaper Engine Service		正在...	自动	本地系...
WarpJITSvc	Provides a JIT out of process service for WARP when running wit...		手动(触...	本地服...
Web 帐户管理器	Web 帐户管理器利用此服务实现应用和服务的单一登录。	正在...	手动	本地系...
WebClient	使基于 Windows 的程序能创建、访问和修改基于 Internet 的文件。...		手动(触...	本地服...
Wifi AutoInstall Service		正在...	自动	本地系...
Windows Audio	管理基于 Windows 的程序的音频。如果此服务被停止，音频设备和效...	正在...	自动	本地服...
Windows Audio Endpoint Builder	管理 Windows 音频服务的音频设备。如果停止了此服务，音频设备和...	正在...	自动	本地系...
Windows Biometric Service	Windows 生物识别服务使客户端应用程序能够捕获、比较、操纵和存...	正在...	手动(触...	本地系...
Windows Camera Frame Server	允许多个客户端从相机设备访问视频帧。		手动(触...	本地系...

扩展　标准

图 2-14　新增加的系统服务

2.2.3　下载 CentOS Stream 9 的 ISO 文件

在后续的实验中，我们将使用 CentOS Stream 9 构建虚拟化宿主机，所以需要从阿里云的开源镜像站(https://mirrors.aliyun.com/centos-stream/9-stream/BaseOS/x86_64/iso/)上下载 CentOS Stream 9 的 ISO 文件，例如 CentOS-Stream-9-latest-x86_64-dvd1.iso，文件约为 8.7GB，如图 2-15 所示。

File Name	File Size	Date
Parent directory/	-	-
CentOS-Stream-9-latest-x86_64-dvd1.iso	8.7 GB	2023-8-19 19:23

图 2-15　阿里云镜像站

2.2.4　创建新虚拟机

图 2-16 所示为 Vmware Workstation 16 Pro 的主界面，可以看到其中共有三个选项，分别是：

- 创建新的虚拟机：即创建一个全新的虚拟机。
- 打开虚拟机：即打开一个现有的虚拟机。
- 连接远程服务器：即连接到虚拟机的服务器上。

这里因为需要先创建一个全新的虚拟机，所以选择“创建新的虚拟机”。

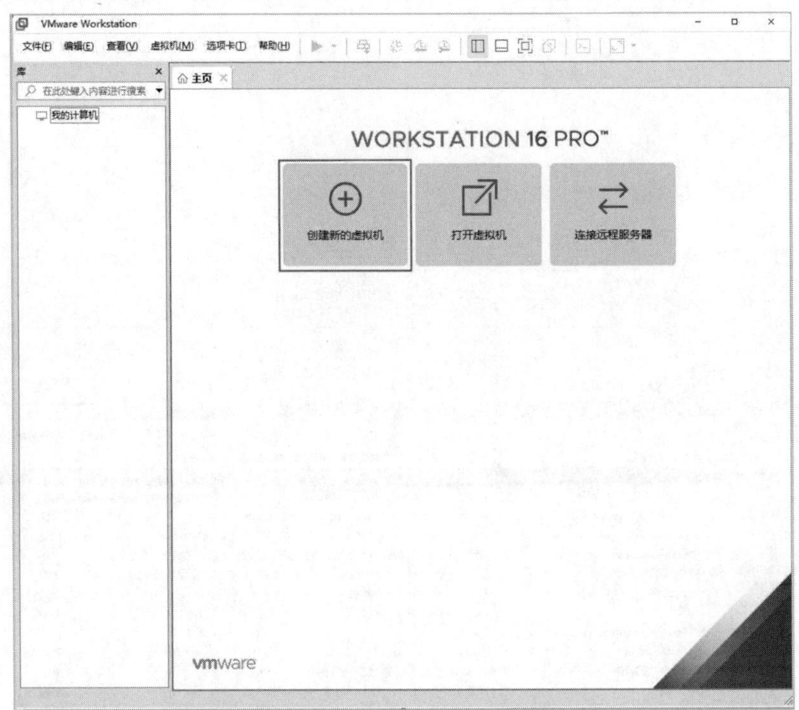

图 2-16　VMware Workstation 16 Pro 主界面

单击“创建新的虚拟机”后系统会跳转到新建虚拟机向导，可以看到其中有两个选项，一个是“典型”，另一个是“自定义”。需要注意的是，一般在安装 Windows 系统的时候才会选择“典型”，安装其他系统时则选择“自定义”。所以这里我们选择“自定义”，如图 2-17 所示。

单击“下一步”按钮后程序会让我们选择硬件兼容性，也就是说对产品的兼容性进行选择，除了产品的兼容性，还有一些硬件限制。因为所安装的宿主机没有太大的空间，所以这里选择 Workstation 16.2.x，如图 2-18 所示。

图 2-17　新建虚拟机向导

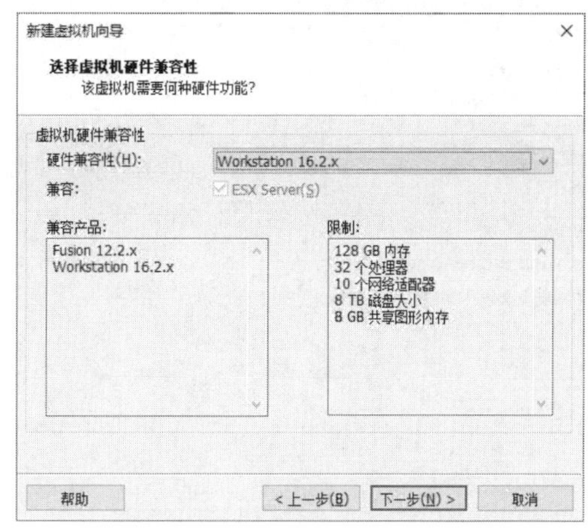

图 2-18　硬件兼容性

在新建虚拟机向导中选择客户机操作系统的来源时，选中“稍后安装操作系统”便会创建一个有空白硬盘的虚拟机，如图 2-19 所示。这里选择“稍后安装操作系统”。

接下来要选择对应的客户机操作系统，因为 CentOS 属于 Linux 的操作系统，所以我们选择 Linux。前面所下载的系统为“CentOS Stream 9”，在下面版本处并没有相关的选项，所以要查找和系统相近的选项，这里找到“CentOS 8 64 位”，然后单击“下一步”，如图 2-20 所示。

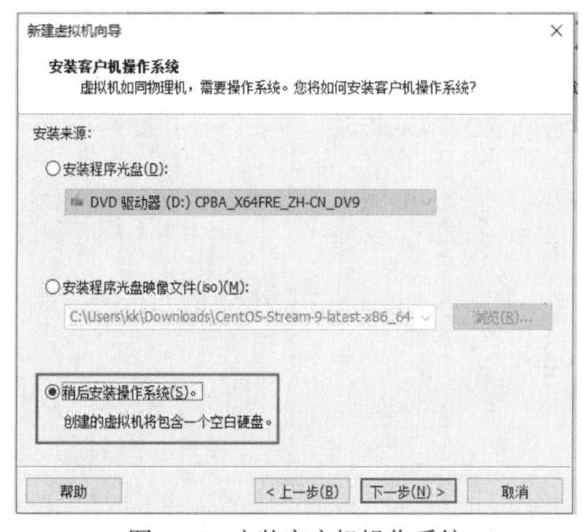

图 2-19　安装客户机操作系统

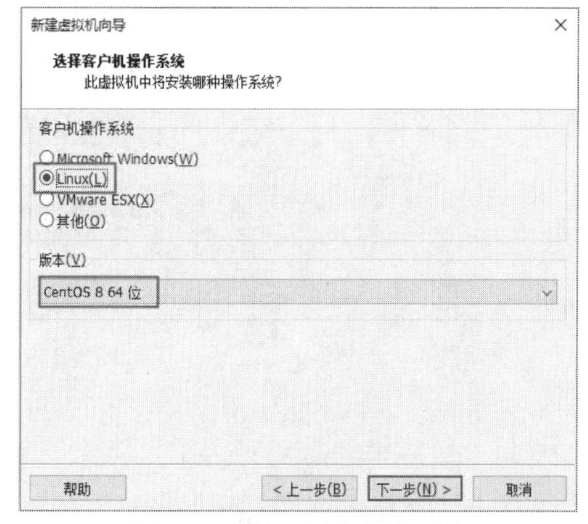

图 2-20　选择客户机操作系统

接下来对虚拟机进行命名(图 2-21)，命名时使用一些有意义的名称，这里我们命名为“KVM1”，对“位置”尽量选择非 C 盘的其他盘，因为虚拟机占用的磁盘空间较大，如果都放在 C 盘会造成系统的卡顿。但如果只有 C 盘，则不进行更改。

接下来是关于虚拟机处理器配置的向导，可以指定具体有几个 CPU，每个 CPU 有几个内核，需要注意的是处理器内核总数不能超过计算器的内核总数。因为只是做实验使用，够用就行，所以选择 1 个处理器，每个处理器的内核数量有 2 个，也就是我们通常所说的单 CPU 双核心，如图 2-22 所示。

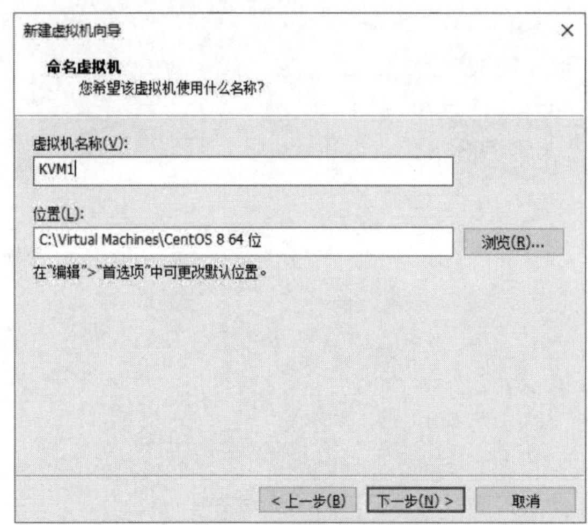

图 2-21　虚拟机位置及命名

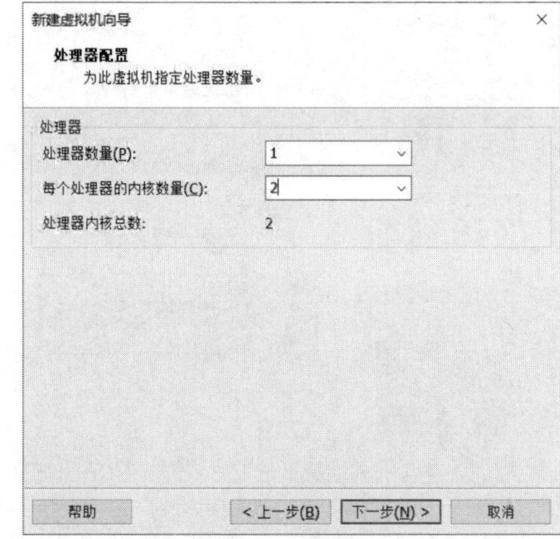

图 2-22　配置处理器

下面就是虚拟机内存的设置了，和 CPU 一样，所设置的资源不能超过物理机的资源数，推荐使用 4GB 的内存进行实验，可以直接拖动左侧的滑块，也可以直接在右侧的输入框内输入 4096，也就是 4GB，如图 2-23 所示。

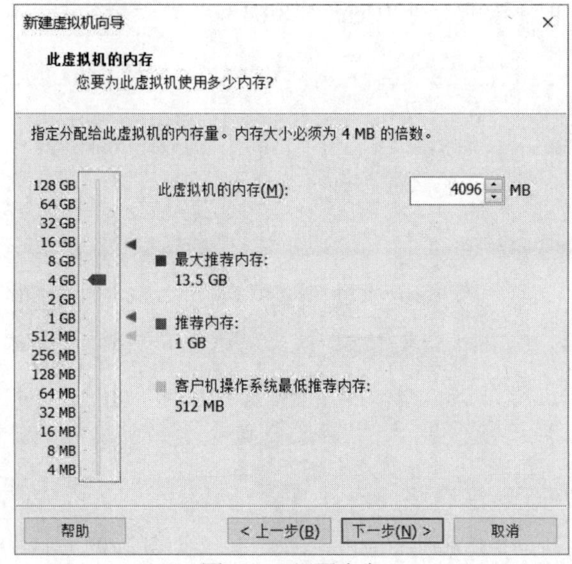

图 2-23　配置内存

上面是关于计算机的一些硬件配置操作，接下来进行网络的配置。其中有四个选项，分别是“使用桥接网络”“使用网络地址转换”“使用仅主机模式网络”和“不使用网络连接”，这里选择“使用网络地址转换”的方式进行联网，如图 2-24 所示。

接下来要选择 I/O 设备的控制器，需要注意的是 BusLogic 是比较老的技术。I/O 控制器的性能较差，只对一些老的系统有效，LSI Logic SAS 的性能比 LSI Logic 好一些，而 LSI Logic 已经够用了，所以没必要选择 SAS，这里选择 LSI Logic，如图 2-25 所示。

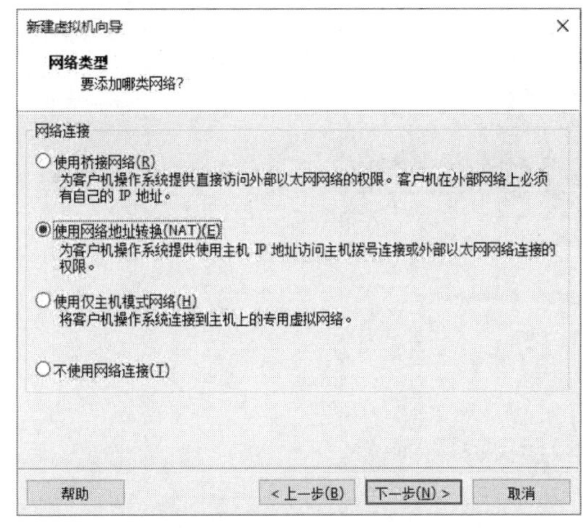

图 2-24　配置网络类型

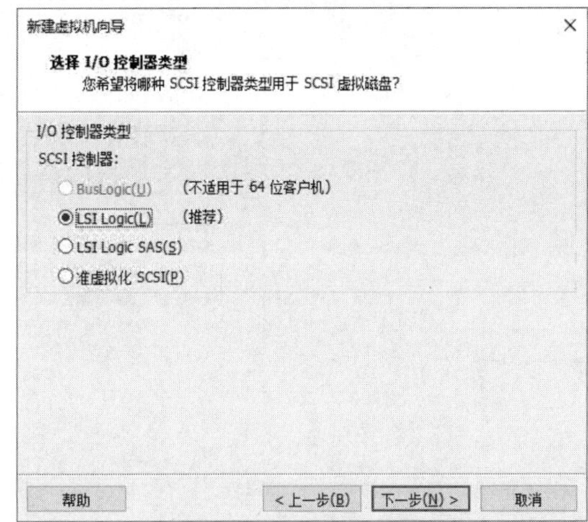

图 2-25　配置 I/O 控制器

单击“下一步”按钮，进入选择磁盘类型界面，如图 2-26 所示。采用默认设置，即选择 NVMe，单击“下一步”按钮开始选择磁盘，所打开的界面中有三个选项，分别是：

- 创建新虚拟磁盘：将虚拟机的文件都写在一个或多个文件内，将主机上的文件视为虚拟机的硬盘。虚拟磁盘文件可以在不同宿主机上进行复制等操作。
- 使用现有虚拟磁盘：之前创建过虚拟磁盘，想要继续使用之前的虚拟磁盘。
- 使用物理磁盘：直接将文件写入到本地文件系统。

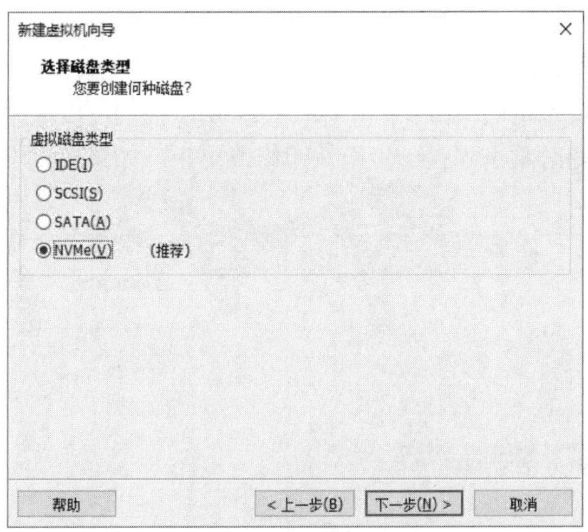

图 2-26　选择磁盘类型

可以看出“创建新虚拟磁盘”会创建出一个全新的磁盘文件，这里为了减少实验的干扰，我们选择“创建新虚拟磁盘”，并单击“下一步”按钮，如图 2-27 所示。

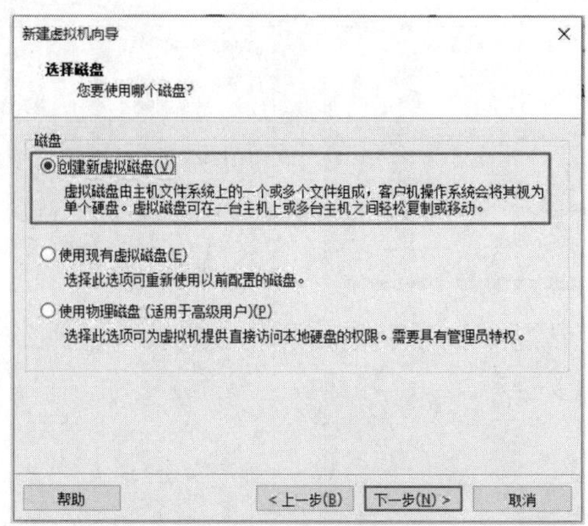

图 2-27　创建新虚拟磁盘

在打开的“指定磁盘容量”界面(图 2-28)中将“最大磁盘大小”设置为 60GB，接着选择“将虚拟磁盘存储为单个文件”，这样做的目的是为了后面需要移植时能够更方便地找到磁盘文件。单击“下一步”按钮。

注意

“立即分配所有磁盘空间”选项如果被选上了，系统会立即占用物理机的空间，所以这里需要检查下是否选中了该项。如果选了该项，需要手动取消选择该项。

在打开的界面中为磁盘文件命名，如图 2-29 所示。继续单击“下一步”按钮，系统会在打开的界面中将所设置的虚拟机的配置罗列出来(图 2-30)，我们可以根据列表清单来进行核对。如果发现有问题，可以随时单击“上一步”按钮修改更正；如果核对没问题，则单击“完成”按钮完成虚拟机的创建。

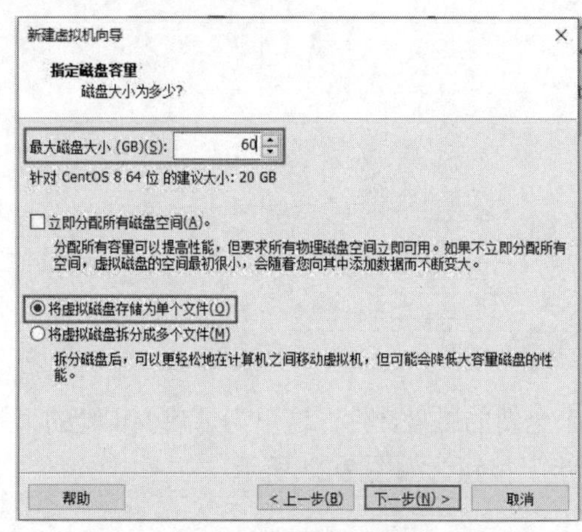

图 2-28　配置磁盘容量

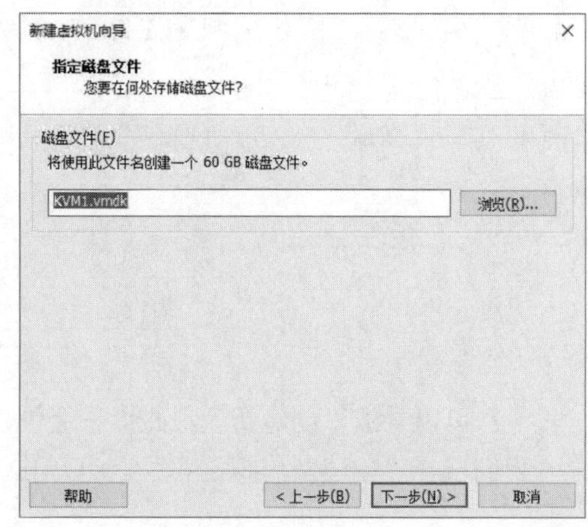

图 2-29　为磁盘文件命名

图 2-30　完成虚拟机的创建

创建好虚拟机后，但是虚拟机的 VT 还没打开，这里可以在 VMware 的主界面中单击“编辑虚拟机设置”，如图 2-31 所示。当弹出来虚拟机设置对话框的时候，选择“处理器”然后就可以看到 VT 的开关，如图 2-32 所示，要确保“虚拟化 Intel VT-x/EPT 或 AMD -V/RVI”复选框处于选中状态。

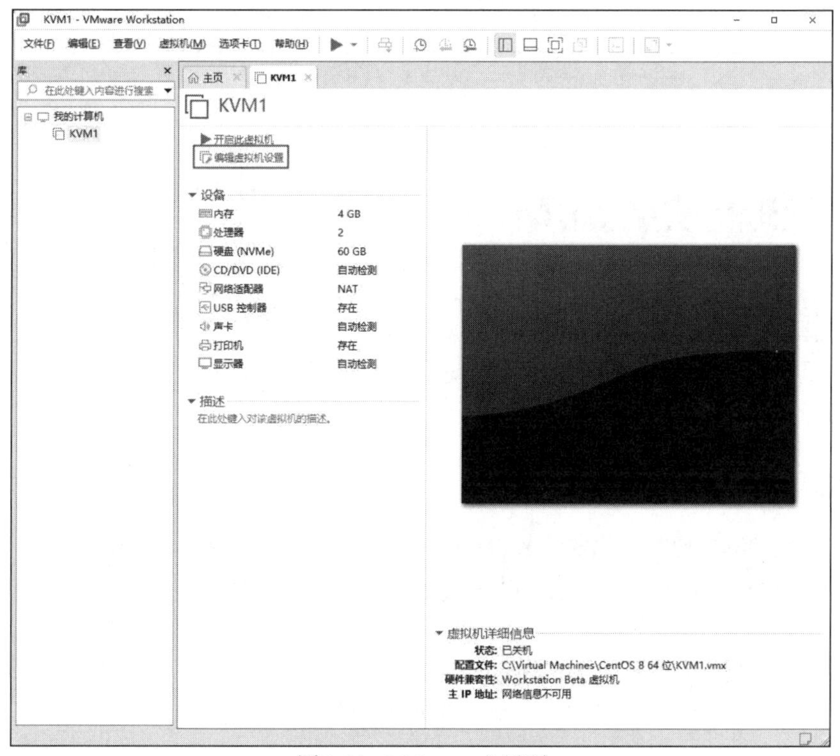

图 2-31　VMware 主界面

虚拟机创建完成后需要打开虚拟机的虚拟化功能，如果虚拟化功能没打开，在后面安装 KVM 虚拟机的时候就会出现异常，默认情况下虚拟化功能是处于打开状态的，但需要进行确认。

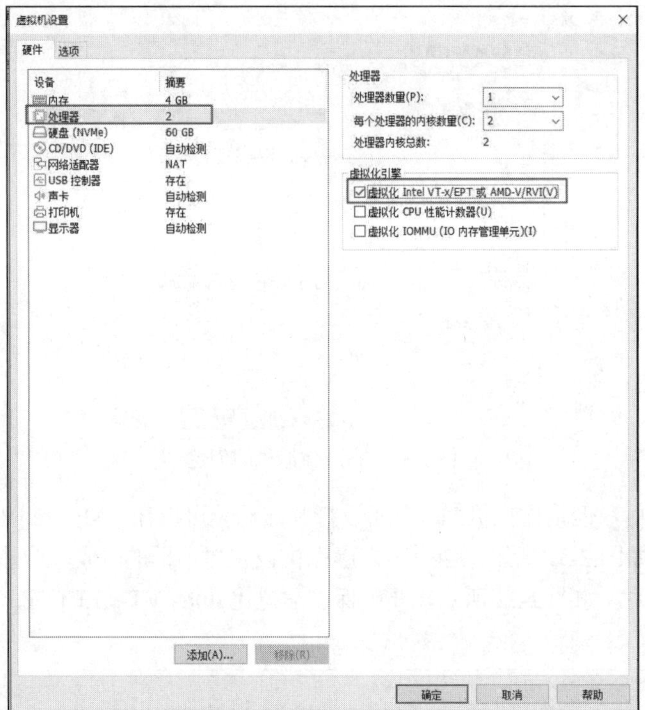

图 2-32　开启虚拟机的 VT

2.3 图形化安装 KVM

上节讲解了如何创建虚拟机，接下来为虚拟机安装操作系统，有了操作系统，系统才能正常运行。这里需要两台虚拟机，我们已经创建了一台虚拟机，另外一台可以直接创建，也可以直接克隆虚拟机。

2.3.1 克隆虚拟机

本章将分别讲解如何图形化安装 KVM 和最小化安装 KVM，所以需要两台虚拟机。另外一台虚拟机可以参考 2.2.4 节创建，也可以在原有的基础上进行克隆，克隆得到的虚拟机也是独立存在的。

打开 VMware 的库，可以看到刚刚创建的虚拟机，在此虚拟机上直接单击鼠标右键，在弹出的菜单栏中选择“管理”，找到“克隆”选项，如图 2-33 所示。

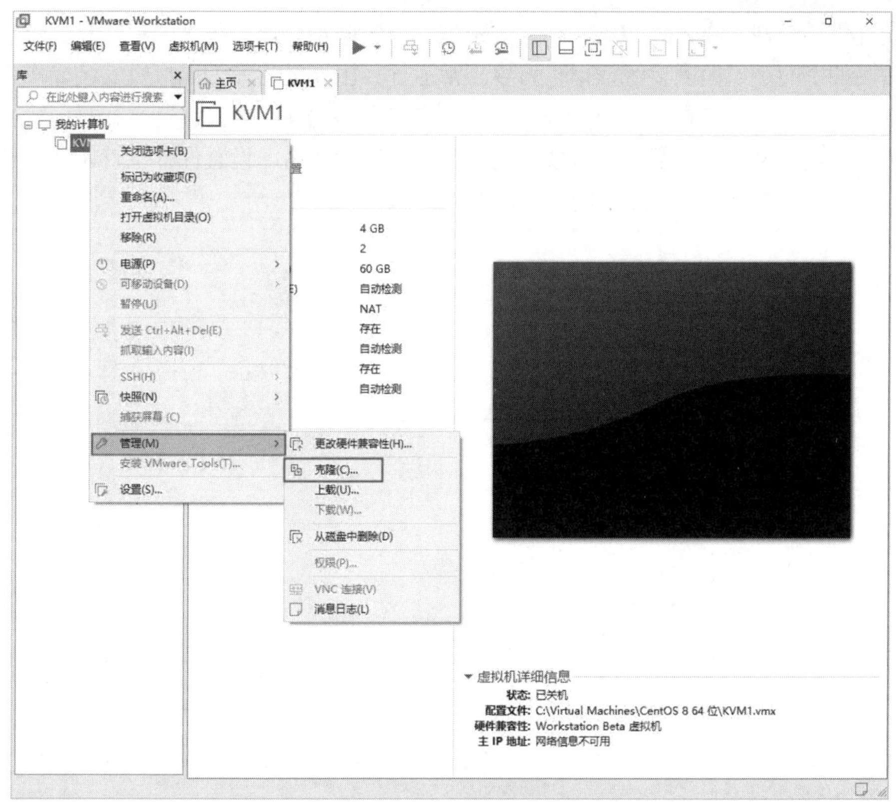

图 2-33　克隆虚拟机

选择“克隆”选项，弹出“克隆虚拟机向导”窗口，如图 2-34 所示，由于在安装 VMware Workstation 的时候已经接受过用户许可协议，所以这里并不会出现接受许可协议的页面。

单击“下一页”之后系统会让我们选择克隆源(图 2-35)，这里有两个选项，第一个是“虚拟机中的当前状态”，第二个是“现有快照”，当前并没有对虚拟机创建过快照，所以这时第二个选项处于不可用的状态，如果创建过快照，且虚拟机处于关闭状态，“现有快照”选项就会变成可用状态，这里直接选择“虚拟机中的当前状态”，并单击“下一页”按钮，进入下一个界面。

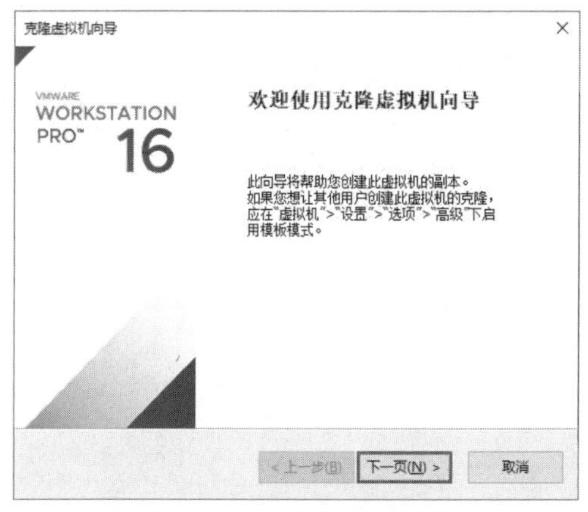

图 2-34　克隆虚拟机向导

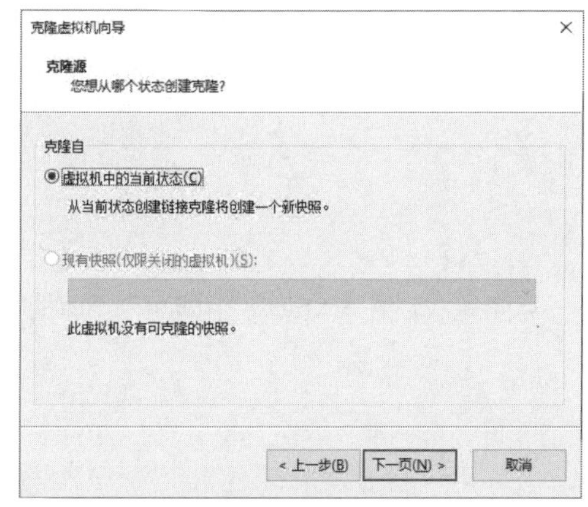

图 2-35　克隆源

系统提供了两种克隆方式(图 2-36)，分别是“创建链接克隆”和“创建完整克隆”，链接克隆所占用的空间小，但需要源虚拟机可以正常运行。由于之前未为虚拟机安装操作系统，所以直接选择“创建完整克隆”，并单击“下一页”继续克隆。

选择完克隆类型之后，需要对克隆出来的虚拟机进行命名，设置保存位置。为了和原来的虚拟机做区分，这里把虚拟机名称修改为 KVM2，如图 2-37 所示。

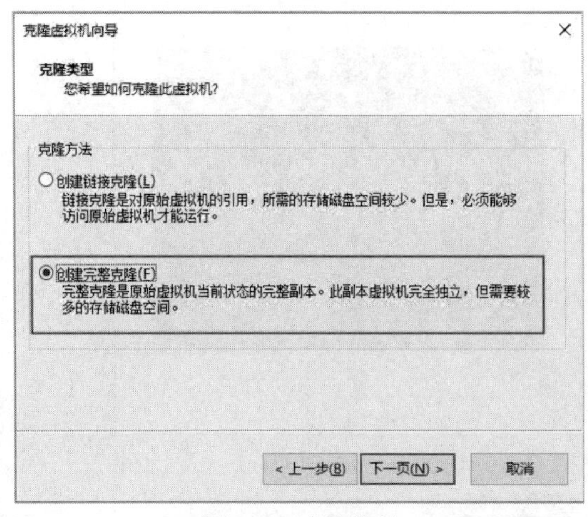

图 2-36　克隆类型

图 2-37　新虚拟机位置及命名

单击“完成”按钮等待克隆完成即可。克隆结束后将会看到如图 2-38 所示的克隆完成界面，单击“关闭”将向导关闭。

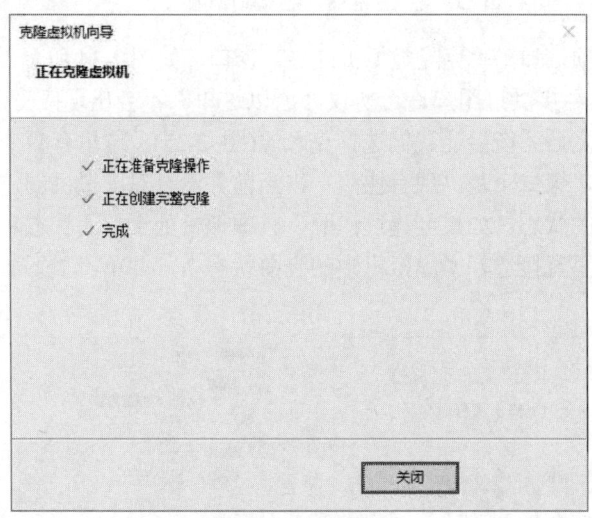

图 2-38　克隆完成

克隆完成后在 VMware 中即可看到刚刚克隆出来的虚拟机，如图 2-39 所示。

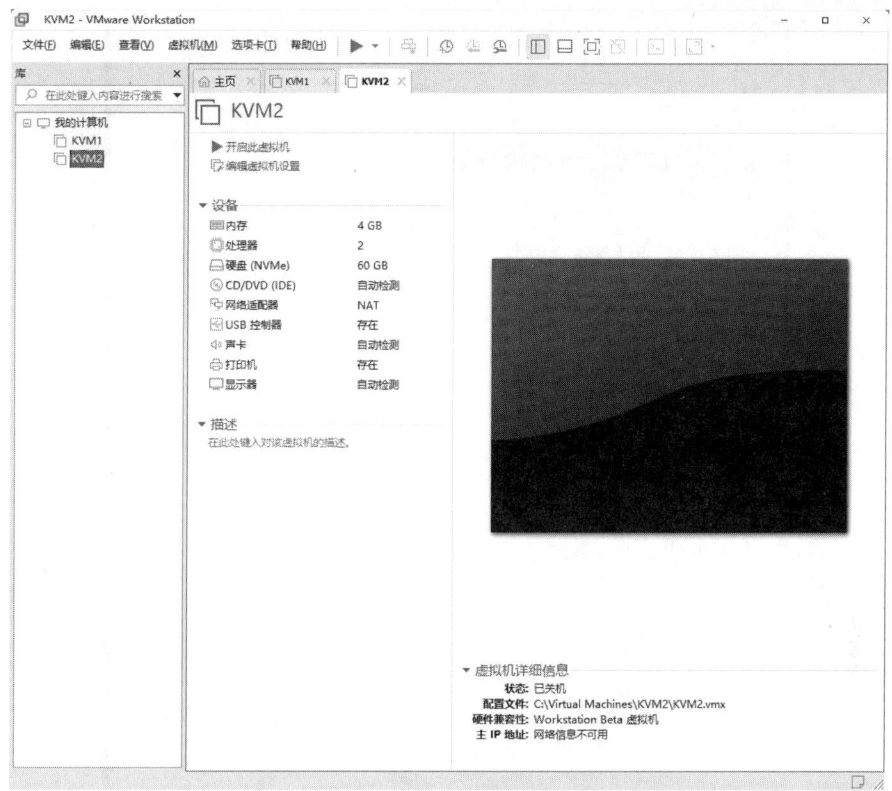

图 2-39　克隆完成后的 VMware 主界面

2.3.2　为虚拟机安装操作系统

虚拟机已经创建好了，下一步就要给虚拟机安装操作系统。回到“虚拟机设置”界面，找到“CD/DVD(IDE)”选项，在右侧选择“使用 ISO 映像文件”，找到我们下载的 CentOS-Stream-9-latest-x86_64-dvd1.iso 文件，如图 2-40 所示。随后选择“开启此虚拟机”，根据提示进行安装即可。

放入 ISO 镜像(也被称为映像)文件之后启动虚拟机，就可以看到如图 2-41 所示的 CentOS Stream 9 安装向导的界面。其中一共有三个选项，分别是：

- Install CentOS Stream 9：立即安装 CentOS Stream 9 操作系统。
- Test this media & install CentOS Stream 9：检测镜像文件，并安装 CentOS Stream 9 操作系统。
- Troubleshooting：修复故障。

这里选择第一个“Install CentOS Stream 9”选项来进行系统的安装。

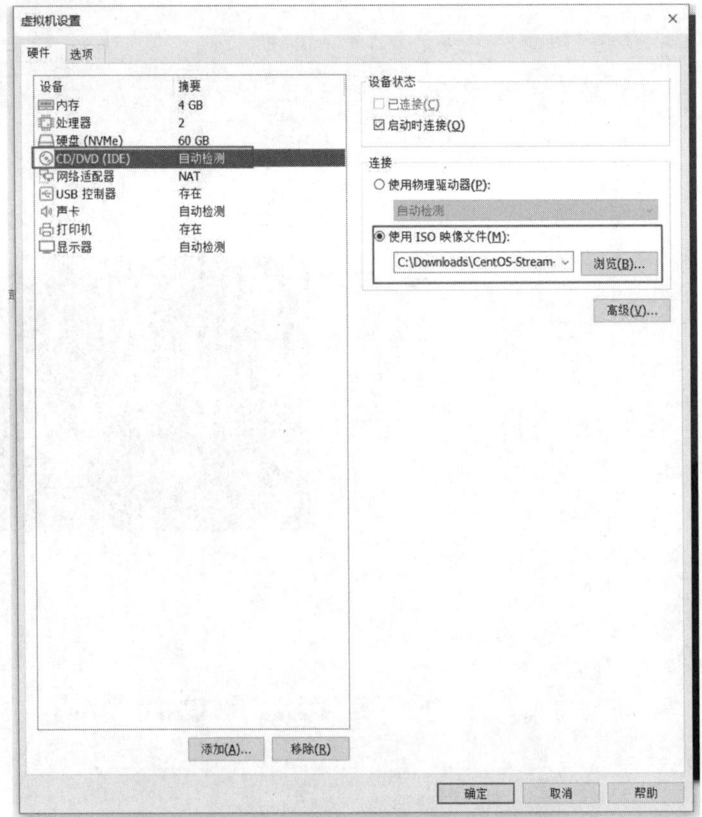

图 2-40　编辑 CD/DVD 并使用 ISO 镜像文件

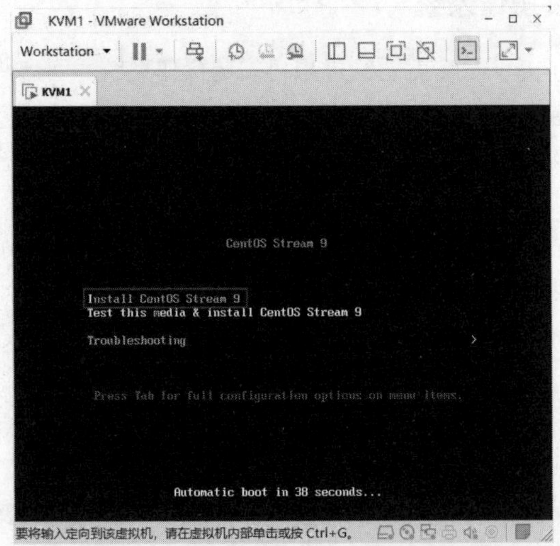

图 2-41　CentOS Stream 9 安装向导

进入选择语言界面，正常选择“中文”→“简体中文”就行，如果找不到“中文”选项，可以直接在下面的输入框内输入“中文”来进行搜索，如图 2-42 所示。

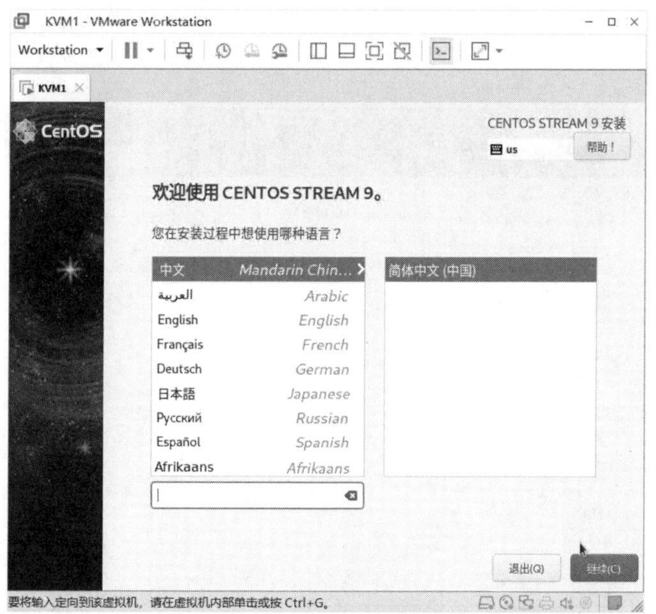

图 2-42　语言选择

选择完语言之后，需要进一步配置安装信息。需要选择安装的目的地、安装的软件，以及需要给 root 用户设置密码，设置完这 3 项之后才可以进行下一步，如图 2-43 所示。

图 2-43　配置安装信息

接下来打开的图 2-44 是配置安装磁盘的界面，也就是选择“安装目的地”打开的页面，这里需要勾选上我们的磁盘，然后单击左上角的“完成”按钮。如果没有选择磁盘，会发现“完成”按钮是灰色无法单击的。

图 2-44　配置安装磁盘

图 2-45 是通过选择“软件选择”所打开的页面，在这个界面中选择要安装的系统类型，CentOS 中集成了各种环境，这里需要根据自己的需求进行选择，如图 2-45 所示，基本环境包括以下几项：

- Server with GUI(带图形用户界面的服务器)
- Server(服务器)

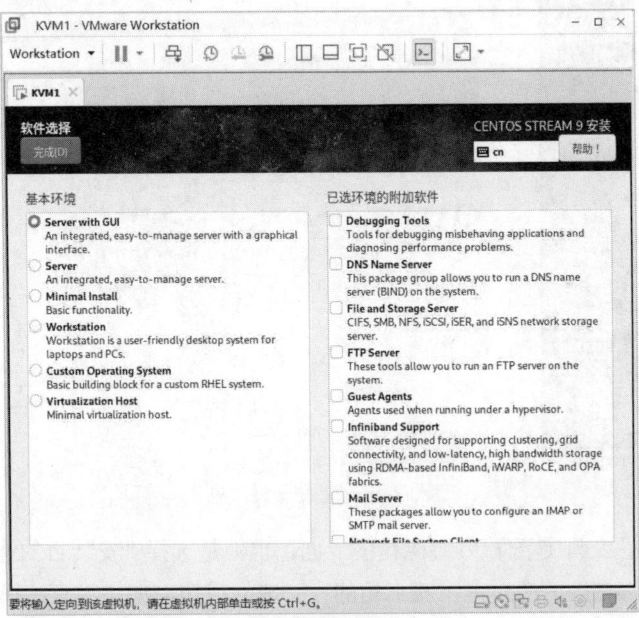

图 2-45　软件选择

- Minimal Install(最小安装)

- Workstation(工作站)
- Custom Operation System(定制操作系统)
- Virtualization Host(虚拟化主机)

接下来先讲解图形化。先安装图形化系统，选择“Server with GUI”选项，并单击左上角的“完成”即可。

设置用户 root 密码时应遵循密码的设置规则，如果必须设置弱强度的密码，也可以直接设置。需要注意的是这里的完成需要单击两次才能确认，如图 2-46 所示，且需要选中“允许 root 用户使用密码进行 SSH 登录”选项。

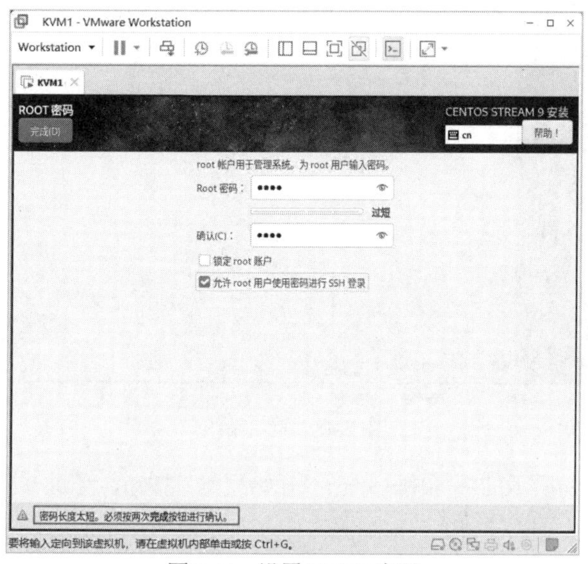

图 2-46　设置 ROOT 密码

把所有需要设置的内容都设置好之后，系统中的“开始安装”按钮才会变为可用状态，如图 2-47 所示。如果“开始安装”按钮一直不可用，请检查是否有其他设置没有设置好。

图 2-47　开始安装

开始安装后只需要耐心等待程序的安装即可，待到程序安装成功后还需要对系统进行重启，安装完成的界面如图 2-48 所示，单击“重启系统”按钮，系统启动完成后就可以使用 CentOS Stream 9 系统了。

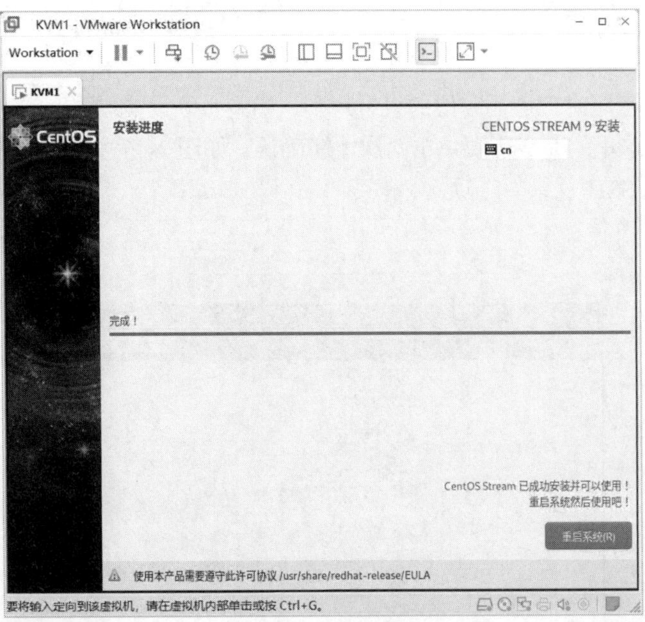

图 2-48　重启系统

在打开的如图 2-49 所示的界面中，单击“开始配置”按钮，经过一些简单的配置之后系统就安装好了，接下来就可以对系统进行拍照操作，对系统拍摄快照有助于在系统有问题的时候进行恢复操作。

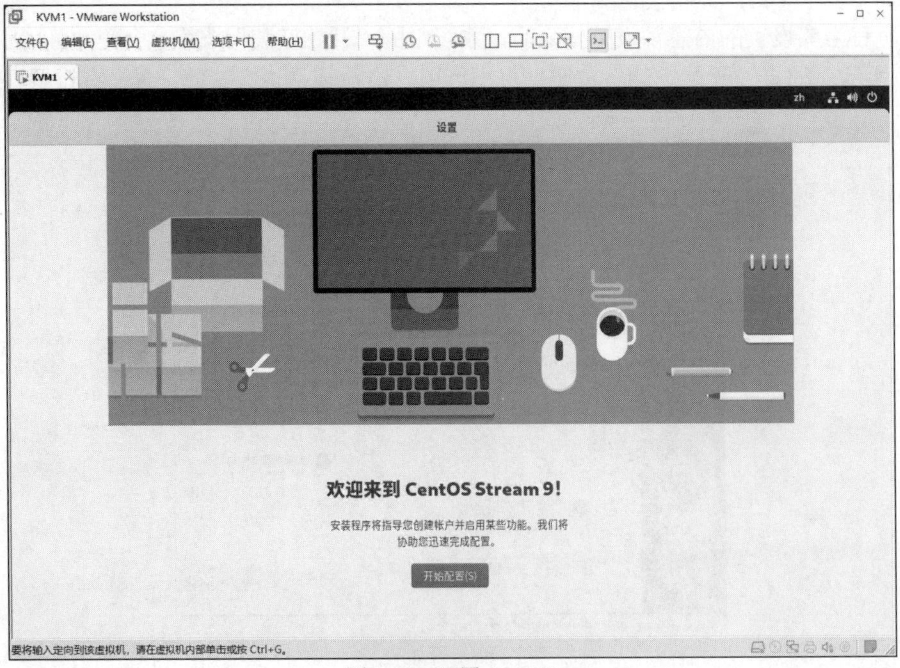

图 2-49　配置 CentOS

2.3.3 图形化下安装 KVM

上一节已经把虚拟机的操作系统安装好了，本节将在上一节的基础上安装 KVM，如果上一节的内容没有安装好请先安装系统。

在 CentOS 中打开终端，在终端中输入以下命令来验证安装结果。

```
# cat ~/anaconda-ks.cfg
    …
    %packages
    @^graphical-server-environment
    %end
    …
```

注意

后面将统一使用 root 账号进行登录，但在实际项目中，应尽量避免使用 root 账号。

可以看到，在安装系统的时候没有选择安装其他软件，所以系统中的软件包非常少。要安装 KVM，还需要安装 virtualization-host-environment、remote-system- management 和 virtualization-platform。安装命令如下：

```
# dnf group install -y virtualization-hypervisor virtualization-platform virtualization-tools
```

也可以写成如下格式：

```
# dnf group install virtualization-hypervisor
# dnf group install virtualization-platform
# dnf group install virtualization-tools
```

安装好后，可以通过如下三条命令验证是否安装成功。

```
# dnf group info virtualization-host-environment
    上次元数据过期检查：0:08:19 前，执行于 2022 年 10 月 16 日 星期日 08 时 12 分 16 秒。
    环境组：Virtualization Host
    描述：Minimal virtualization host.
    必选软件包组：
    Base
    Core
    Standard
    Virtualization Hypervisor
    Virtualization Tools
    可选软件包组：
    Debugging Tools
    Network File System Client
    Remote Management for Linux
    Virtualization Platform
# dnf group info remote-system-management
```

```
    上次元数据过期检查：0:06:09 前，执行于 2022 年 10 月 16 日 星期日 08 时 12 分 16 秒。
    组：Remote Management for Linux
    描述：Remote management interface for Red Hat Enterprise Linux.
    默认的软件包：
    cockpit
    net-snmp
    net-snmp-utils
    rhel-system-roles
    sblim-cmpi-base
    tog-pegasus
    wsmancli
    可选的软件包：
    openwsman-server
    sblim-indication_helper
    sblim-sfcb
    sblim-wbemcli
# dnf group info virtualization-platform
    上次元数据过期检查：0:07:02 前，执行于 2022 年 10 月 16 日 星期日 08 时 12 分 16 秒。
    组：Virtualization Platform
    描述：Provides an interface for accessing and controlling virtualized guests and containers.
    必要的软件包：
    libvirt
    libvirt-client
    virt-who
    可选的软件包：
    fence-virtd-libvirt
    fence-virtd-multicast
    fence-virtd-serial
```

CentOS 9 的安装程序还会启动一个名为 libvirtd 的服务，但这个服务并不会自动启动，需要通过手动的方式进行启动。我们可以通过如下命令查看 libvirtd 的启动状态，当看到 Active 的状态为 inactive(dead)时，表示我们的服务当前处于不活跃状态，也就是没有启动。

```
# systemctl status libvirtd
○ libvirtd.service - Virtualization daemon
     Loaded: loaded (/usr/lib/systemd/system/libvirtd.service; disabled; vendor preset>
     Active: inactive (dead)        #不活跃状态
     TriggeredBy: ○ libvirtd-tcp.socket
                  ○ libvirtd-tls.socket
                  ○ libvirtd.socket
                  ○ libvirtd-admin.socket
                  ○ libvirtd-ro.socket
     Docs: man:libvirtd(8)
           https://libvirt.org
lines 1-10/10 (END)
```

可以通过以下命令启动 libvirtd，输入以下命令后如果没有报错则表示运行成功，我们可以通过重新运行第二行代码来验证运行状态。

```
# systemctl start libvirtd
# systemctl status libvirtd

# systemctl status libvirtd
● libvirtd.service - Virtualization daemon
    Loaded: loaded (/usr/lib/systemd/system/libvirtd.service; disabled; vendor preset>
    Active: active (running) since Sun 2022-10-16 08:23:46 CST; 3s ago
    #可以看到已经是活跃状态了，并且已经把启动的时间和运行时长也告诉了我们。
    TriggeredBy: ○ libvirtd-tcp.socket
                 ○ libvirtd-tls.socket
                 ● libvirtd.socket
                 ● libvirtd-admin.socket
                 ● libvirtd-ro.socket
   Docs: man:libvirtd(8)
         https://libvirt.org
   Main PID: 33742 (libvirtd)
   Tasks: 21 (limit: 32768)
   Memory: 21.1M
   CPU: 377ms
…
```

从上面的运行提示中可以看出，目前 libvirtd 服务已经在正常运行了，接下来就可以尝试运行 virt-manager 命令来唤起“虚拟系统管理器”窗口，当输入 virt-manager 的时候会发现系统找不到命令，但是紧接着系统会询问我们是否安装对应的包，这里直接输入“Y”就可以进行安装，整体的流程如下：

```
# virt-manager
   bash: virt-manager: command not found...
   Install package 'virt-manager' to provide command 'virt-manager'? [N/Y] Y



    * Waiting in queue...
    * Loading list of packages....
   The following packages have to be installed:
    gtk-vnc2-1.3.0-1.el9.x86_64 A GTK3 widget for VNC clients
    gvnc-1.3.0-1.el9.x86_64   a GObject for VNC connections
    libburn-1.5.4-4.el9.x86_64Library for reading, mastering and writing   optical discs
    libisoburn-1.5.4-4.el9.x86_64Library to enable creation and expansion of ISO-9660 filesystems
    libisofs-1.5.4-4.el9.x86_64Library to create ISO 9660 disk images
    libvirt-daemon-kvm-8.9.0-2.el9.x86_64 Server side daemon & driver required  to run KVM guests
    libvirt-glib-4.0.0-3.el9.x86_64  libvirt glib integration for events
    python3-argcomplete-1.12.0-5.el9.noarch  Bash tab completion for argparse
    virt-manager-4.1.0-1.el9.noarchDesktop tool for managing virtual machines via libvirt
    virt-manager-common-4.1.0-1.el9.noarch  Common files used by the different Virtual Machine
    Manager interfaces
    xorriso-1.5.4-4.el9.x86_64ISO-9660 and Rock Ridge image manipulation tool
   Proceed with changes? [N/Y] Y
```

```
* Waiting in queue...
* Waiting for authentication...
* Waiting in queue...
* Loading list of packages....
* Downloading packages...
* Requesting data...
* Testing changes...
* Installing packages...
```

系统运行完以上代码之后，会弹出“虚拟系统管理器”窗口，看到窗口正常弹出，也就意味着已经安装完成，具体界面如图 2-50 所示。

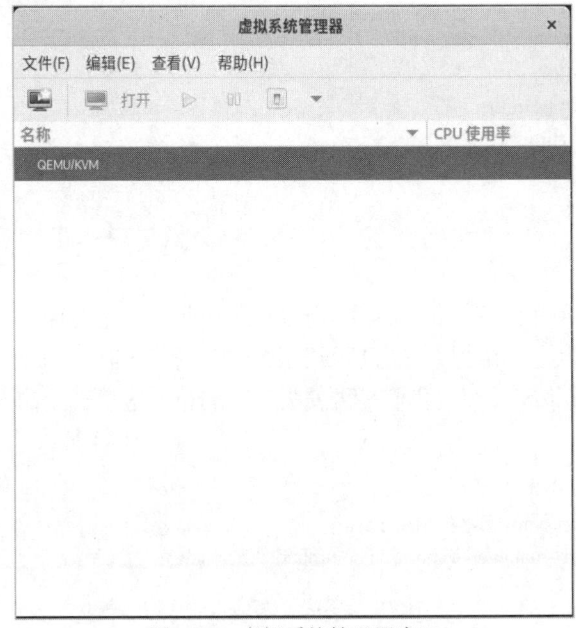

图 2-50　虚拟系统管理器窗口

2.4 系统最小化安装 KVM

上一节讲到了如何在图形化界面下安装 KVM，本节继续讲解在最小化系统下安装 KVM 的过程。在此之前需要确保已创建了虚拟机，接下来介绍在最小化系统下应该如何安装 KVM。

2.4.1　最小化系统的安装

要想安装 KVM，首先还是需要先将虚拟机安装在操作系统中，本节所使用的硬件配置和前面安装图形界面时所使用的配置一样，配置清单如下：

- CPU：双核 CPU
- 内存：4GB
- 硬盘：60GB

安装步骤和安装图形化一致，唯一不同的是在选择软件的时候所选择的是 Minimal Install 选项，如图 2-51 和图 2-52 所示。

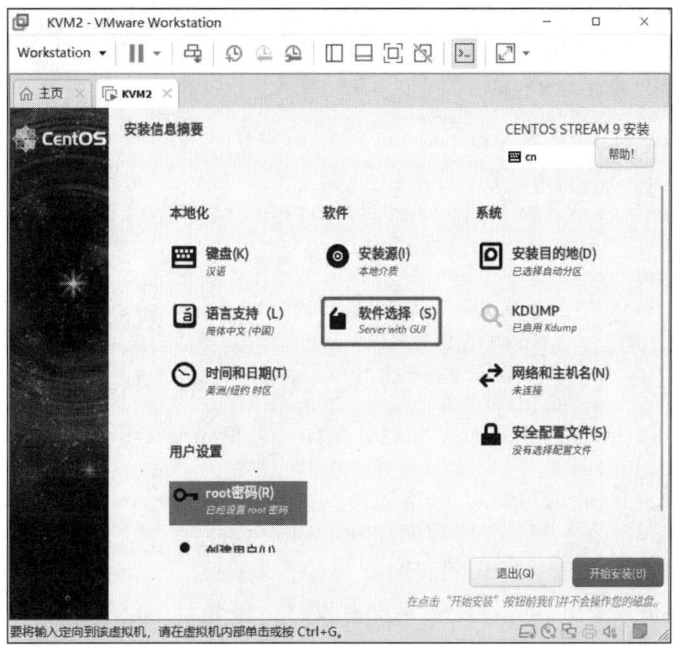

图 2-51 软件选择

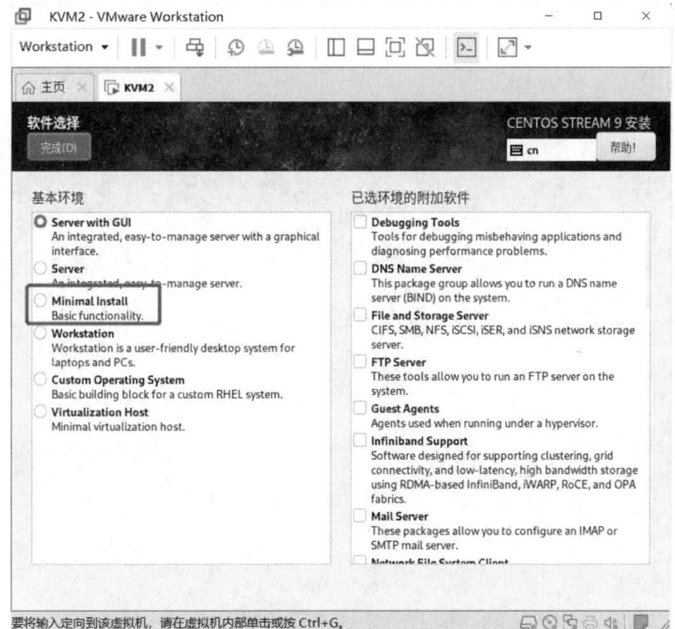

图 2-52 选择 Minimal Install 选项

设置好之后只需要安静地等待系统安装完成即可。

系统完成安装之后，先查看虚拟机的 IP 地址，以便后面连接虚拟机的时候使用。由于选择了 Minimal Install 选项，所以无法通过 ifconfig 命令查看虚拟机的 IP 地址，而是需要通过 ip a

命令查看虚拟机的 IP 地址。

```
# ip a
    1: lo: <LOOPBACK,UP,LOWER_UP> mtu 65536 qdisc noqueue state UNKNOWN group default    qlen 1000
    link/loopback 00:00:00:00:00:00 brd 00:00:00:00:00:00
    inet 127.0.0.1/8 scope host lo
        valid_lft forever preferred_lft forever
    inet6 ::1/128 scope host
        valid_lft forever preferred_lft forever
    2: ens160: <BROADCAST,MULTICAST,UP,LOWER_UP> mtu 1500 qdisc mq state UP group default qlen 1000
    link/ether 00:0c:29:da:3a:6e brd ff:ff:ff:ff:ff:ff
    altname enp3s0
    inet 192.168.100.13/24 brd 192.168.100.255 scope global dynamic noprefixroute ens160
        valid_lft 5497960sec preferred_lft 5497960sec
    inet6 fe80::20c:29ff:feda:3a6e/64 scope link noprefixroute
        valid_lft forever preferred_lft forever
    3: virbr0: <NO-CARRIER,BROADCAST,MULTICAST,UP> mtu 1500 qdisc noqueue state DOWN group
      default qlen 1000
    link/ether 52:54:00:ad:99:30 brd ff:ff:ff:ff:ff:ff
    inet 192.168.122.1/24 brd 192.168.122.255 scope global virbr0
        valid_lft forever preferred_lft forever
```

从上面的结果可以看出，在 ens160 后面有我们的 IP 地址，这里先记录下来。

接下来需要下载 Xmanager 软件，这里要注意的是需要使用 Xmanager 和 Xshell 两个软件，两个软件都必须下载后才能正常使用。打开 Xmanager-Passive，以接收 SSH 转发至本地的图形界面信息，我们可以在“开始”菜单栏里面直接搜索，如图 2-53 所示，搜索到之后直接单击“打开”，软件会自动运行并放置在后台。

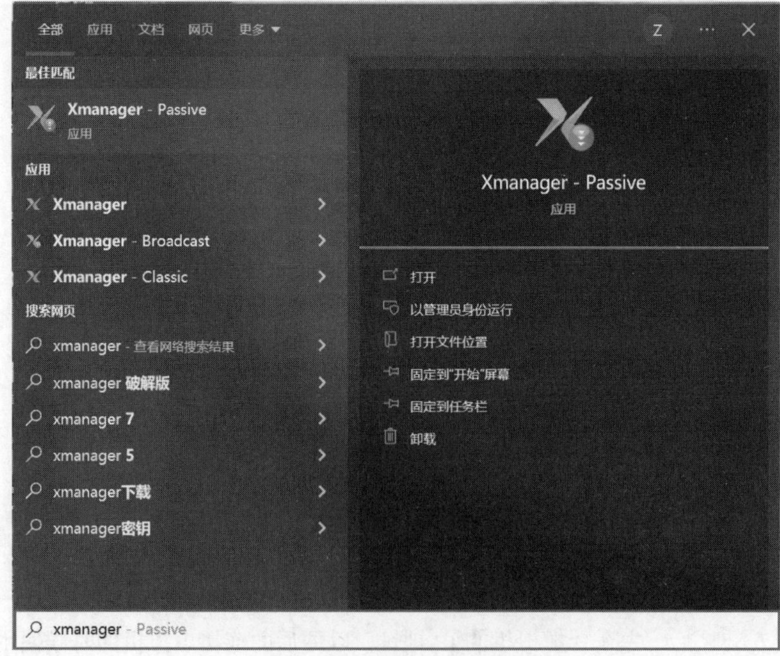

图 2-53　Xmanager-Passive

另外还需要打开 Xshell 的 X11 转移功能，选择 Xshell 的“文件”→“默认会话属性”菜单，打开“默认会话属性”窗口，在“默认会话属性”窗口中选择“X11 转移”中的 Xmanager，具体操作可以参考图 2-54 和图 2-55。

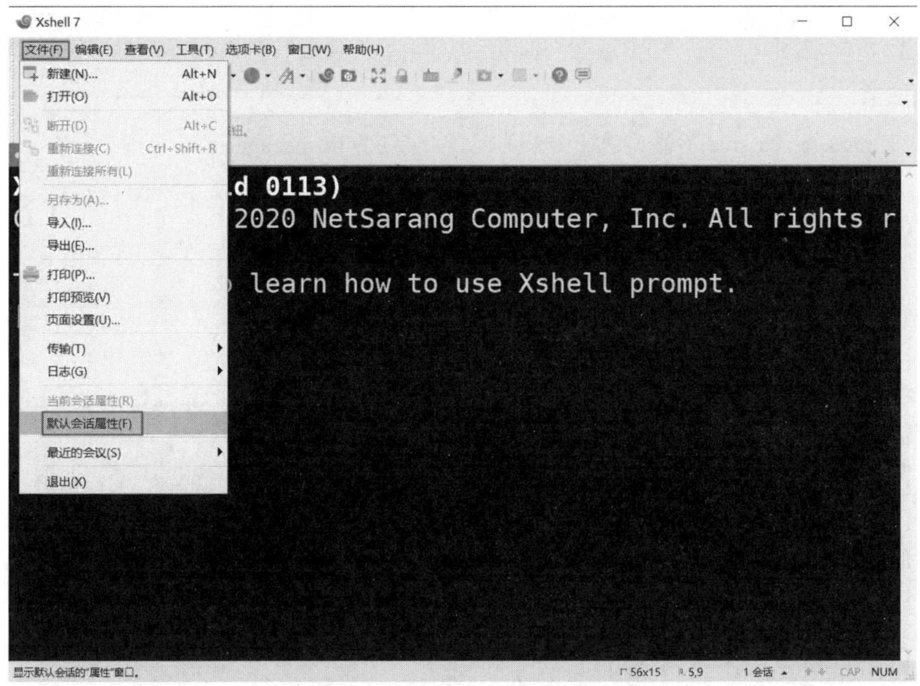

图 2-54　Xshell 默认会话属性的位置

图 2-55　Xshell 的 X11 转移

设置好之后，就可以使用 Xshell 进行连接了，这里可以选择在 Xmanager 上打开服务器的远程连接，也可以选择直接在Xshell中打开远程连接，推荐大家直接在Xmanager里面打开Xshell会话。

Xshell 可以使用创建的新连接，也可以在地址栏输入服务器的地址进行连接，具体的操作方法不再赘述，可以参照图 2-56 来输入。

注意

Xmanager-Passive 不能关闭，如果关闭，会导致 virt-manager 的连接断开。

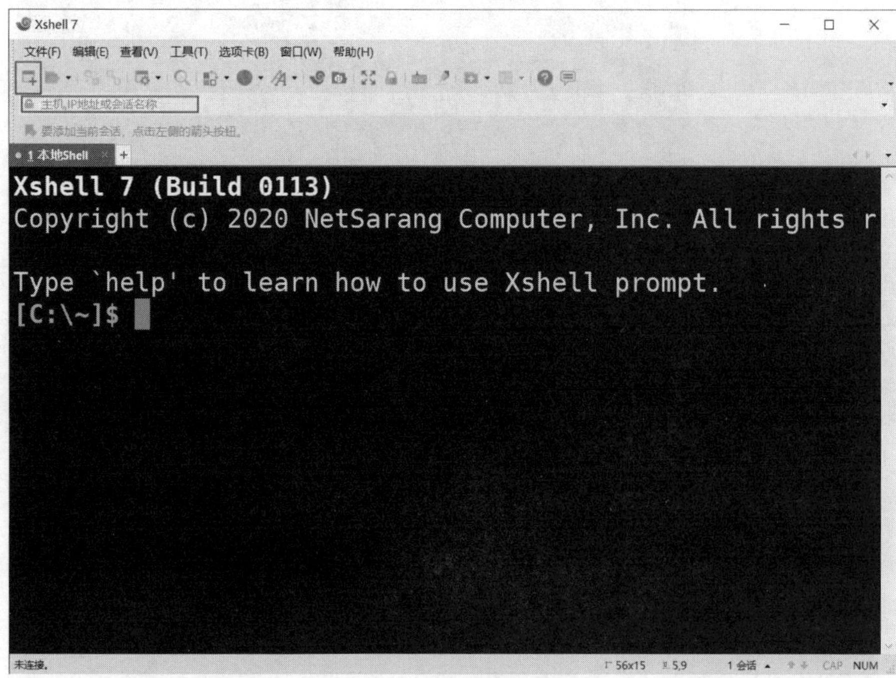

图 2-56　Xshell 连接

接下来在 Xshell 中输入以下命令并运行：

- #安装 KVM 图形化管理工具。

```
# dnf -y install virt-manager openssh-askpass
```

- #安装 X11 图形化管理工具。

```
#   dnf -y install xorg-x11-server-utils xorg-x11-utils xorg-x11-xauth xorg-x11-xinit.x86_64
```

- #解决使用时出现乱码的情况。

```
#   dnf -y groupinstall -y "Fonts"
#   dnf group install -y "Fonts"
```

- #开启 ssh 的 X11 转发功能。

```
# vi /etc/ssh/sshd_config
    X11Forwarding yes
```

```
# systemctl restart sshd
```

以上命令运行完成后，就可以使用 virt-manager &命令来运行 KVM 图形化管理界面。

```
# virt-manager &
```

当看到图形化管理窗口正常弹出后即表示当前 KVM 已经安装好了。

2.5 KVM 管理工具

前面讲解了关于 KVM 的各种安装方式，本节将介绍 KVM 的管理工具。KVM 的管理工具主要有 libvirt、virsh、virt-manager、virt-viewer 等。

2.5.1 libvirt

libvirt 用于管理虚拟化平台的开源 API、后台程序。它可以用于管理 KVM、Xen、VMware ESX、QEMU 和其他虚拟化技术，这些 API 在云计算的解决方案中被广泛使用。

libvirt 是提供了一个方便的方式来管理虚拟机和其他虚拟化功能的软件的集合，如存储和网络接口管理。这些软件包括一个 API 库、一个守护进程(libvirtd)和一个命令行实用程序(virsh)。

libvirt 的首要目标是能够管理多个不同的虚拟化供应商/虚拟机管理程序，以提供一个单一的方式。例如，命令“virsh 列表”等都可以用于任何支持现有的虚拟机管理程序列表(如 KVM、Xen、VMware ESX 等)，而不需要学习管理程序特定的工具。

libvirt 所支持的虚拟化平台主要有 LXC、OpenVZ、kernel-based virtual machine/QEMU (KVM)、Xen、VirtualBox、VMware ESX and GSX、Hyper-V 等。

2.5.2 virsh

virsh 是使用 libvirt management API 构建的管理工具，是 KVM 虚拟机常用的管理工具。virsh 是 virtualization shell 的编写。virsh 有命令模式和交互模式，如果直接在 virsh 后面添加的参数是命令模式，如果直接写 virsh，就会进入交互模式。

```
virsh --help                 #查看命令帮忙
virsh list                   #显示正在运行或挂起的虚拟机
virsh list --all             #显示所有的虚拟机
virsh start domain           #启动虚拟机
virsh shutdown domain        #关闭虚拟机
virsh destroy domain         #虚拟机强制断电(易丢失数据，慎用)
virsh suspend domain         #挂起虚拟机
virsh resume domain          #恢复挂起的虚拟机
virsh reboot domain          #重启虚拟机
virsh undefine domain        #删除虚拟机(慎用)，会删除默认路径下(/etc/libvirt/qemu/)的虚拟机
                              配置文件
virsh define domain.xml      #用于虚拟机迁移后的导入，domain.xml 为虚拟机的配置文件(只要
                              有虚拟机的磁盘文件加配置文件，那么虚拟机便可以迁移到任何
```

```
                                     地方并导入)
virsh dominfo domain                 #查看虚拟机的配置信息
virsh domiflist domain               #查看虚拟机网卡的配置信息
virsh domblklist domain              #查看该虚拟机的磁盘位置
virsh edit domain                    #修改虚拟机的 xml 配置文件/etc/libvirt/qemu/domain.xml(带语法检查)
virsh dumpxml domain                 #查看虚拟机的当前配置
virsh dumpxml vm-node1 > vm-node1.bak.xml  #备份 vm-node1 虚拟机的 xml 文件，原文件默认路径为
                                           /etc/libvirt/qemu/vm-node1.xml
virsh autostart domain               #KVM 物理机开机自启动虚拟机，配置后会在此目录生成配置软连
                                      接文件/etc/libvirt/qemu/autostart/vm-node1.xml。所以，本质就是将虚
                                      拟机的配置文件移到该目录下，手动移动也生效。如果物理机断电，
                                      上电后，希望 KVM 虚拟机能够立即自启动
virsh autostart --disable domain     #取消开机自启动
virsh vncdisplay domain              #查看虚拟机 VNC 端口号(第一台默认为 5901)
virsh domblklist domain              #列出虚拟机的所有块设备
virsh desc template1                 #查看虚拟机的描述信息
virsh domrename domain_old domain_new#修改虚拟机名(关机状态下)
virsh pool-list
virsh pool-list --all
virsh net-list --all
virsh net-list
virsh net-edit <网卡>                #编辑网络文件，即/etc/libvirt/qemu/networks/xxx.xml 文件
virsh iface-list                     #物理主机接口列表
virsh nwfilter-list                  #列出网络过滤器
virsh   snapshot-list domain         #列出虚拟机的快照
virsh vol-list <pool>                #列出某地址池的数据卷
virsh nodeinfo                       #查看宿主机的 CPU 信息
```

2.5.3 virt-manager

virt-manager 是一套虚拟机的桌面管理器，与 VMware 的 vCenter 和 xenCenter 类似，工具提供了虚拟机管理的基本功能，如开机、挂起、重启、关机、强制关机/重启、迁移等，并且可以进入虚拟机图形界面进行操作。该工具还可以管理各种存储以及网络运行。

2.5.4 virt-viewer

virt-viewer 是一个最小化的虚拟机图形界面展示工具，支持 VNC 和 SPICE 两种远程协议。

virt-viewer 的用法也很简单，在终端里面输入 virt-manager 命令就可以直接调出对应的操作界面。如图 2-57 所示为选择虚拟机界面。

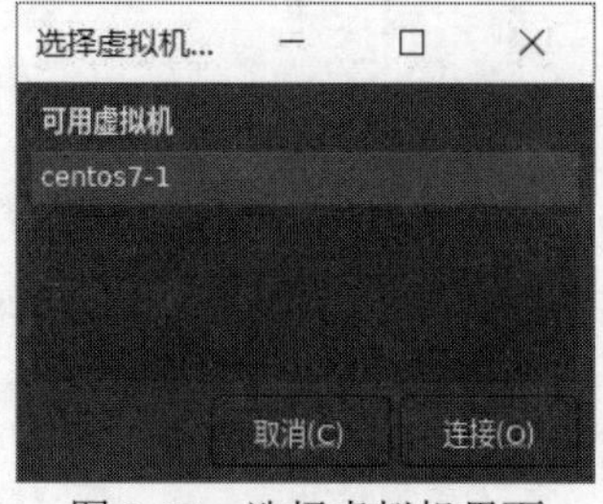

图 2-57　选择虚拟机界面

2.6 习题

1. 动手安装 VMware Workstation 软件。
2. 创建虚拟机。
3. 安装 KVM 虚拟机。
4. 简述最小化系统和图形系统有什么区别。

第3章

创建KVM虚拟机

通过上一章的学习，学员已经掌握了 KVM 的架构，以及如何在图形化系统和最小化系统中安装 KVM，本章我们将着重讲解 KVM 虚拟机的创建。

本章要点

◎ 利用 Virt-Manager 图形化创建 KVM 虚拟机
◎ 利用 virt-install 命令创建虚拟机
◎ 利用 VNC 连接 KVM 虚拟机

3.1 Virt-Manager 图形化创建 KVM 虚拟机

虚拟机管理器(Virt-Manager)是一个管理宿主机和虚拟机的 GUI 管理工具。虽然 RHEL/CentOS 还包含这个软件，会在将来用 Cockpit 替换，可是，目前 Cockpit 功能还不完善，有些功能只能通过 Virt-Manager 或 Virsh 完成，所以我们必须掌握 Virt-Manager 的使用。

3.1.1 创建虚拟机

前面介绍了 Virt-Manager 的基本操作以及如何查看配置，接下来学习如何使用 Virt-Manager 创建虚拟机。

要创建虚拟机需要先在虚拟机上规划出一些位置来存储所对应的文件，这里创建一个 iso 目录、一个 kvm-lvm 目录，其中 iso 目录用来存储 iso 镜像文件，而 kvm-lvm 用来存储虚拟机。

```
# mkdir /iso
# mkdir /kvm-lvm
```

```
# cd /iso
# wget http://mirror.nsc.liu.se/centos-store/7.5.1804/isos/x86_64/CentOS-7-x86_64-DVD-1804.iso
```

下载结束后，还需要再添加一块硬盘来存储虚拟机。打开 Workstation，选择“虚拟机”→“设置”，如图 3-1 所示，之后弹出如图 3-2 所示的页面，我们可以在这个页面中添加硬盘。

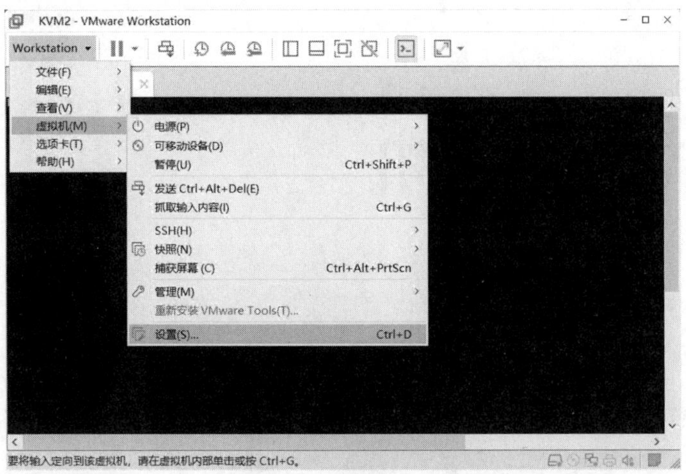

图 3-1　设置虚拟机

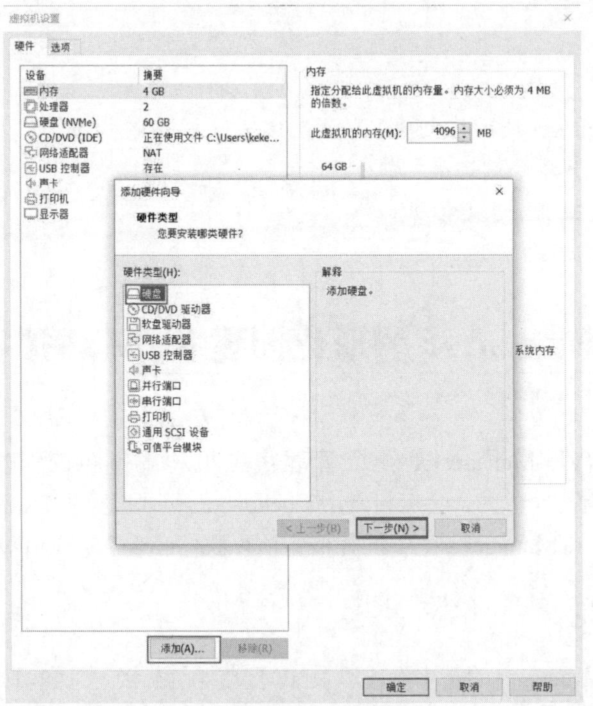

图 3-2　添加硬盘

在如图 3-2 所示的界面中可以看到左侧有具体的虚拟机配置，在下方有“添加”按钮，单击“添加”按钮后会弹出“添加硬件向导”界面，在该界面中不仅能添加硬盘，还可以添加其他硬件，因为当前的需求是添加一块硬盘，所以这里直接选择“硬盘”并单击“下一步”按钮继续完成硬盘的添加。

和创建虚拟机的操作一样，也需要选择磁盘类型，不过不同的地方是如果选择 IDE 或者是 NVMe 类型的磁盘，必须关闭虚拟机，否则没办法添加这两种类型的磁盘。继续使用默认推荐的 SCSI 类型，然后继续单击“下一步”按钮，如图 3-3 所示。

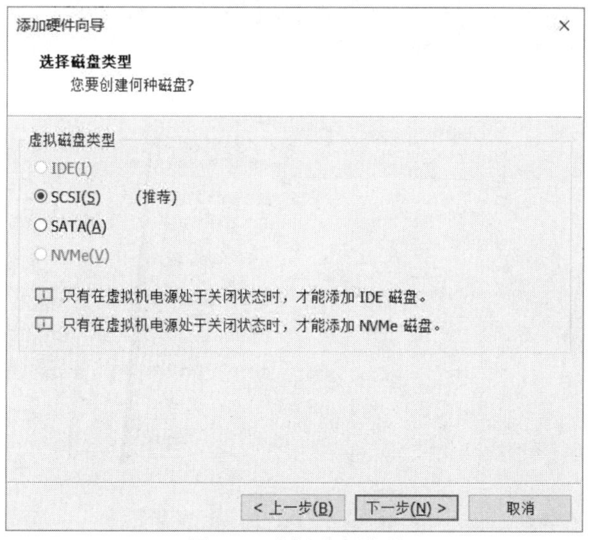

图 3-3 选择磁盘类型

选择完类型之后同样需要设置磁盘的容量及选择文件(图 3-4)，为了区分，把这个磁盘大小设置为 20GB，这样在后面格式化或者分区的时候就可以区分开磁盘，同样需要将磁盘存储为单个文件，“立即分配所有磁盘空间”选项框则应取消选中，继续单击“下一步”。

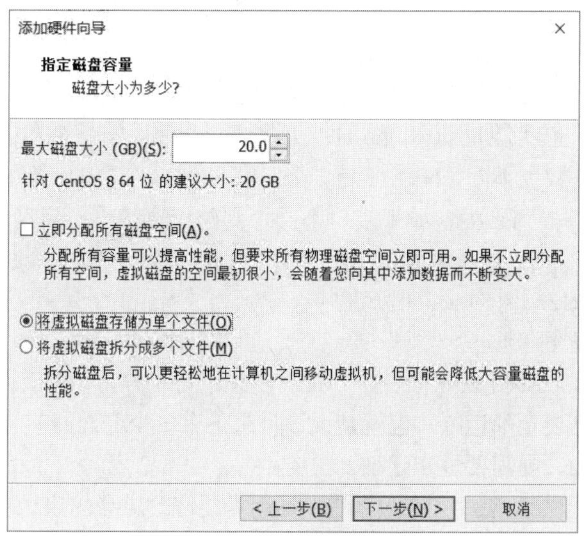

图 3-4 指定磁盘容量

当能在虚拟机设置中看到两块硬盘的时候，就说明硬盘已经成功添加，并被 VMware 识别了，接下来就可以到虚拟机系统里进行设置并使用了，如图 3-5 所示。

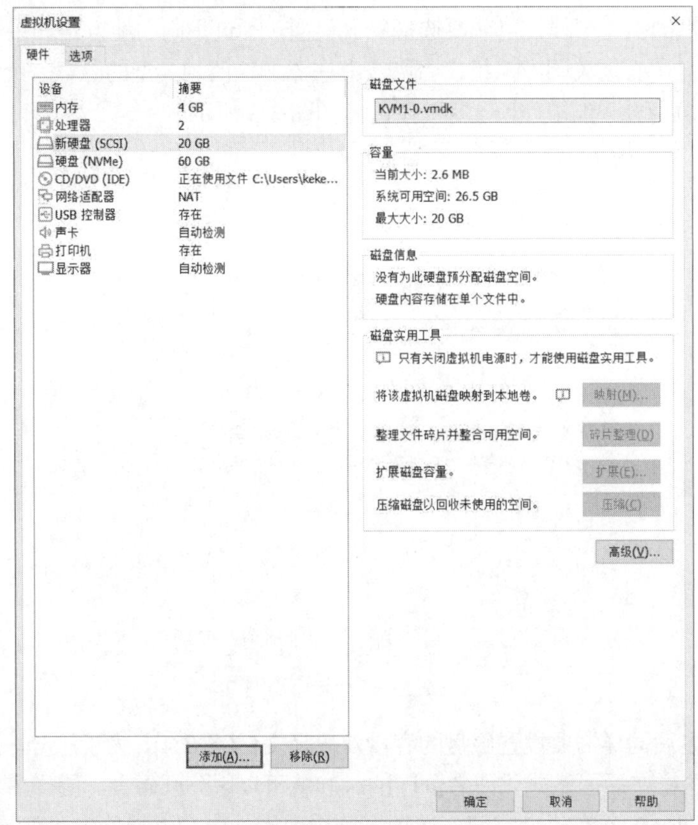

图 3-5　添加后的虚拟机设置

单击“确定”后系统会自动重新启动虚拟机并识别新的磁盘，如果系统没有重新启动，则最好手动重新启动系统。需要注意的是，第一块硬盘使用的是 NVMe 接口，所以在 scsi_host 目录下只有两个目录，也就是 host0 和 host1；第二块硬盘使用的是 SCSI 接口，因为接口不同，scsi_host 目录下的文件数也不同，NVMe 是 2 个，SCSI 是 3 个，IDE 会更多。接口相同的可以使用如下代码让系统扫描到磁盘，如果接口不同，则建议直接重新启动服务器。

```
# echo "- - -" >> /sys/class/scsi_host/host0/scan
# echo "- - -" >> /sys/class/scsi_host/host1/scan
# echo "- - -" >> /sys/class/scsi_host/host2/scan
```

注意，要根据 scsi_host 目录下有多少个 host 目录进行扫描，如果有两个，则运行前两行就行，如果有三个，则需要全部运行，也就是说该目录下有多少个 host 目录就需要运行多少次命令，而后面的 0、1、2，则需要逐步递增。

如果重启后发现无法正常进入系统(图 3-6)，则需要修改磁盘的启动顺序。具体修改方法参考图 3-7 和图 3-8。

```
Network boot from VMware VMXNET3
Copyright (C) 2003-2021  VMware, Inc.
Copyright (C) 1997-2000  Intel Corporation

CLIENT MAC ADDR: 00 0C 29 80 32 29  GUID: 564DEF93-B25B-6946-8A78-EAE4F4803229
PXE-E53: No boot filename received

PXE-M0F: Exiting Intel PXE ROM.
Operating System not found
```

图 3-6　新增磁盘后无法启动

选择红框中所标识的图标，单击这个图标时，VMware 会自动向虚拟机发送一个“Ctrl+Alt+Delete”组合的组合键，也就是我们通常用来调出任务管理器的组合键，在虚拟机出现品牌 logo(图 3-7)的时候点击键盘上行的 F2 键进入到 BIOS 界面。

常用的操作有：方向键的左右键(选择菜单)，方向键的上下键(选择项目)，加号键或者减号键(改变值)，回车键(选择子菜单)，F10 键(保存并退出)。

用左右方向键将选项卡调整到 Boot 页，通过上下方向键选择 Hard Drive，并通过确认键打开对应的子菜单，然后继续使用上下键选择 NVMe(因为本例的系统安装在 NVMe 接口的硬盘上)，通过加号键或者减号键将 NVMe 选项调整至第一个，并按 F10 键保存退出，如图 3-8 所示。

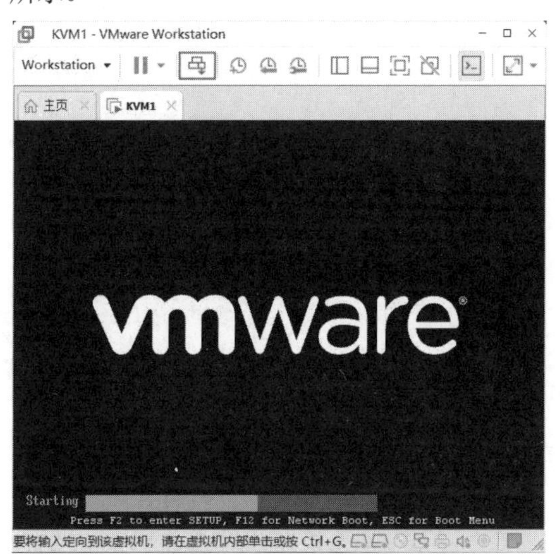

图 3-7　VMware 启动界面

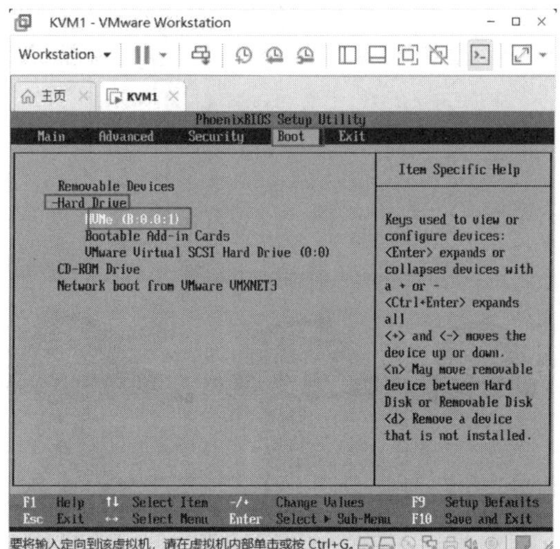

图 3-8　BIOS 设置修改硬盘启动顺序

虚拟机重新启动后就可以正常进入系统，继续完成 KVM 的安装。在开始安装之前，推荐大家把虚拟机拍个快照，以便后面出问题进行恢复时使用。

进入系统后，通过 lsblk 命令可以查看磁盘，这里可以看到 sda 的磁盘，说明系统已经识别到我们的磁盘，列表中的 sda 就是新增加的硬盘，系统盘就是下面的 nvme0n1。

```
# lsblk
NAME          MAJ:MIN RM   SIZE RO TYPE MOUNTPOINTS
sda           8:0      0    20G  0 disk
sr0           11:0     1  1024M  0 rom
nvme0n1       259:0    0    60G  0 disk
├─nvme0n1p1   259:1    0     1G  0 part /boot
```

```
  └─nvme0n1p2       259:2   0       59G    0   part
    ├─cs_192-root   253:0   0       37G    0   lvm   /
    ├─cs_192-swap   253:1   0       3.9G   0-0 lvm   [SWAP]
    └─cs_192-home   253:2   0       18.1G  0   lvm   /home
```

接下来需要对 sda 磁盘进行分区，我们通过 fdisk /dev/sda 进行分区，操作如下：

```
# fdisk /dev/sda

    欢迎使用 fdisk (util-linux 2.37.4)。
    更改将停留在内存中，直到您决定将更改写入磁盘。
    使用写入命令前请三思。

    命令(输入 m 获取帮助)：n
    分区类型
       p   主分区 (0 primary, 0 extended, 4 free)
       e   扩展分区 (逻辑分区容器)
    选择 (默认 p)：p
    分区号 (1-4, 默认 1):
    第一个扇区 (2048-41943039, 默认 2048):
    最后一个扇区，+/-sectors 或 +size{K,M,G,T,P} (2048-41943039, 默认 41943039):

    创建了一个新分区 1，类型为“Linux”，大小为 20 GiB。

    命令(输入 m 获取帮助)：t
    已选择分区 1
    Hex 代码或别名(输入 L 列出所有代码)：8e
    已将分区“Linux”的类型更改为“Linux LVM”。

    命令(输入 m 获取帮助)：w
    分区表已调整。
    将调用 ioctl() 来重新读分区表。
    正在同步磁盘。
```

然后通过 lsblk -l 命令进行查看，可以看出已经有了 sda1 分区，sda1 就是刚创建出来的磁盘分区。

```
# lsblk
    NAME              MAJ:MIN   RM  SIZE    RO TYPE MOUNTPOINTS
    sda                 8:0      0  20G     0 disk
    └─sda1              8:1      0  20G     0 part
    sr0                11:0      1  1024M   0 rom
    nvme0n1           259:0      0  60G     0 disk
    ├─nvme0n1p1       259:1      0  1G      0 part   /boot
    └─nvme0n1p2       259:2      0  59G     0 part
      ├─cs_192-root   253:0      0  37G     0 lvm    /
      ├─cs_192-swap   253:1      0  3.9G    0 lvm    [SWAP]
      └─cs_192-home   253:2      0  18.1G   0 lvm    /home
```

继续创建 lv，并对 lv 进行格式化，代码分别如下：

```
root@192 ~]# pvcreate /dev/sda1
    Physical volume "/dev/sda1" successfully created.
# vgcreate vmvg /dev/sda1
    Volume group "vmvg" successfully created
# lvcreate -L 19.9G -n kvmlv vmvg
    Rounding up size to full physical extent 19.90 GiB
    Logical volume "kvmlv" created.
# mkfs.xfs /dev/vmvg/kvmlv
    meta-data=/dev/vmvg/kvmlv        isize=512    agcount=4, agsize=1304320 blks
             =                       sectsz=512   attr=2, projid32bit=1
             =                       crc=1        finobt=1, sparse=1, rmapbt=0
             =                       reflink=1    bigtime=1 inobtcount=1
    data     =                       bsize=4096   blocks=5217280, imaxpct=25
             =                       sunit=0      swidth=0 blks
    naming   =version 2              bsize=4096   ascii-ci=0, ftype=1
    log      =internal log           bsize=4096   blocks=2560, version=2
             =                       sectsz=512   sunit=0 blks, lazy-count=1
    realtime =none                   extsz=4096   blocks=0, rtextents=0
```

接着就可以直接挂载该分区到/kvm-lvm 目录下，具体命令如下：

```
# mount /dev/vmvg/kvmlv /kvm-lvm/
```

截止到这里，准备工作也就做完了，接下来通过使用 virt-manager 命令唤起 Virt-Manager，并进行虚拟机的创建。

创建虚拟机可以通过菜单栏中的“文件”→“新建虚拟机”来进行，如图 3-9 所示，也可以通过点击新建虚拟机的图标来进行创建，如图 3-10 所示。

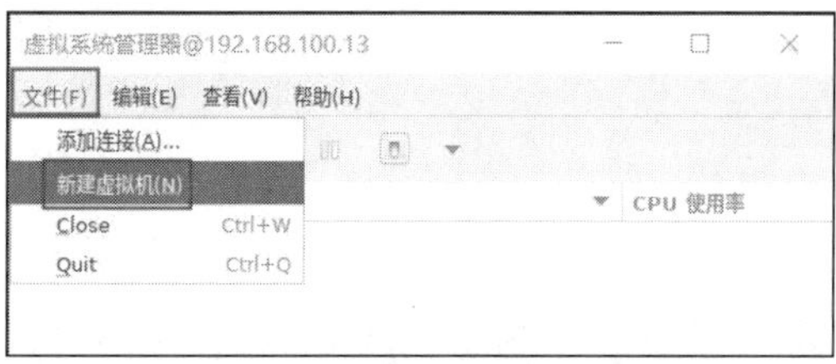

图 3-9　在菜单栏中新建虚拟机

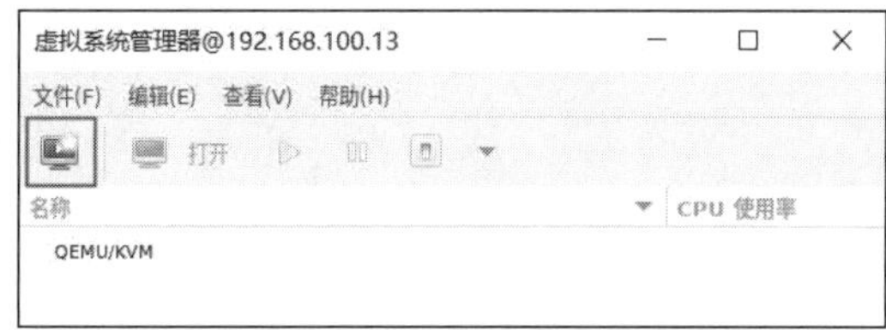

图 3-10　创建新虚拟机的图标

接下来开始创建虚拟机，共分为五个步骤。

第一步：需要在系统给定的四种安装方式中选择一种安装方式进行系统的安装，具体选项说明如下：

(1) 本地安装介质(ISO 镜像或者光驱)

此方法使用 ISO 格式的镜像文件。虽然写有光驱，但是目前还无法通过宿主机上的

CD-ROM 或 DVD-ROM 设备进行安装。

(2) 网络安装(HTTP、HTTPS 或 FTP)

通过保存在 HTTP、HTTPS 或 FTP 服务器上的操作系统安装文件来安装。如果选择此项，还需要提供安装文件的 URL 及内核选项。

(3) 导入现有磁盘镜像

采用与引导执行环境服务器来安装虚拟机。

(4) 手动安装

创建新的虚拟机，并将现有的磁盘镜像(包含预安装的可引导操作系统)导入该虚拟机。

这里选择“本地安装介质(ISO 映像或者光驱)”并单击 Forward 按钮以继续下一步，如图 3-11 所示。

第二步：需要查找和安装介质，如图 3-12 所示。可以自己选择，也可以自动从安装介质中检测，单击“浏览”按钮定位 ISO 介质卷窗口，如图 3-12 和图 3-13 所示。

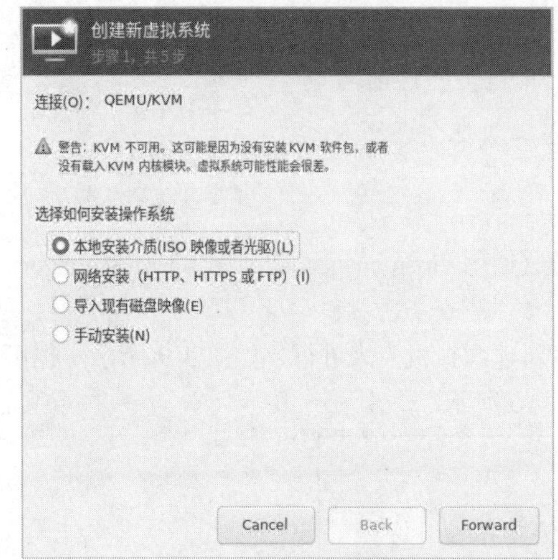

图 3-11　选择安装类型

图 3-12　从本地 ISO 映像安装

图 3-13　选择安装介质

选择完映像文件之后就需要设置存储池了，这里添加两个存储池，即一个 iso 和一个 kvm-lvm。单击左下角的加号即可添加存储池，单击下面的“+”会弹出如图 3-14 所示的创建存储池的界面，这里把第一个存储池的名称命名为 iso，类型选择为 dir：文件系统目录，目标路径为/iso，填写好具体信息之后点击“完成”完成第一个存储池的创建。完成第一个之后继续点击“+”完成第二个存储池的创建，第二个存储池的名称为 vm，类型为 dir：文件系统目录，目标路径为/kvm-lvm，具体的参数如图 3-15 和图 3-16 所示。

图 3-14　选择存储卷

图 3-15　创建 iso 存储池

图 3-16　创建 vm 存储池

第三步：配置内存和 CPU 选项，需要注意的是创建虚拟机和在物理机上创建虚拟机一样，都是不能把内存或 CPU 设置超过虚拟机的数量，这里设置内存为 1024，单位为 M，CPU 的数量设置为 1 个，设置好后点击 Forward 继续下一步，如图 3-17 所示。

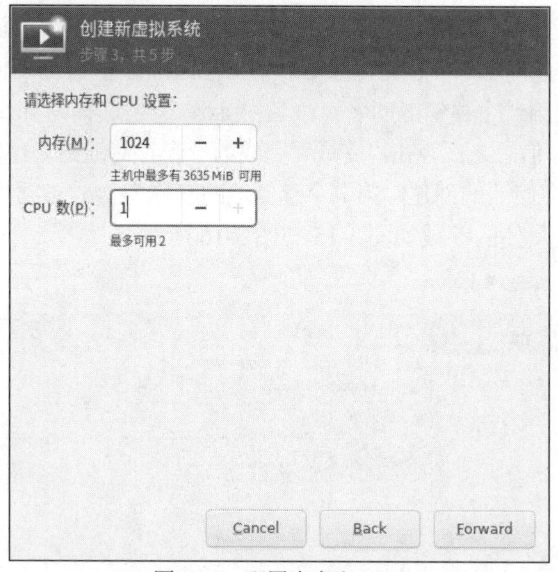

图 3-17　配置内存和 CPU

第四步：配置虚拟机的存储。配置好内存和 CPU 之后就需要对系统的存储进行配置，所谓的存储就是指电脑里的硬盘，在图 3-18 所示的界面中，第一个选项是“为虚拟机启用存储”，必须要选择该项，如果不勾选就等于电脑里没有硬盘，下面是为虚拟机创建磁盘空间的大小，这里我们设置为 10G，设置好之后继续点击 Forward 按钮进行下一步。

第五步：配置虚拟机名称、网络、体系结构和其他硬件设置。单击图 3-18 中的 Forward 按钮，打开如图 3-19 所示的界面，把虚拟机的名称设置为“demon1_centos7”，网络选择为“虚拟网络'default'：NAT”模式，然后点击“完成”即可开始安装操作系统，如图 3-20 所示。

图 3-18　配置虚拟磁盘

图 3-19　安装清单确认

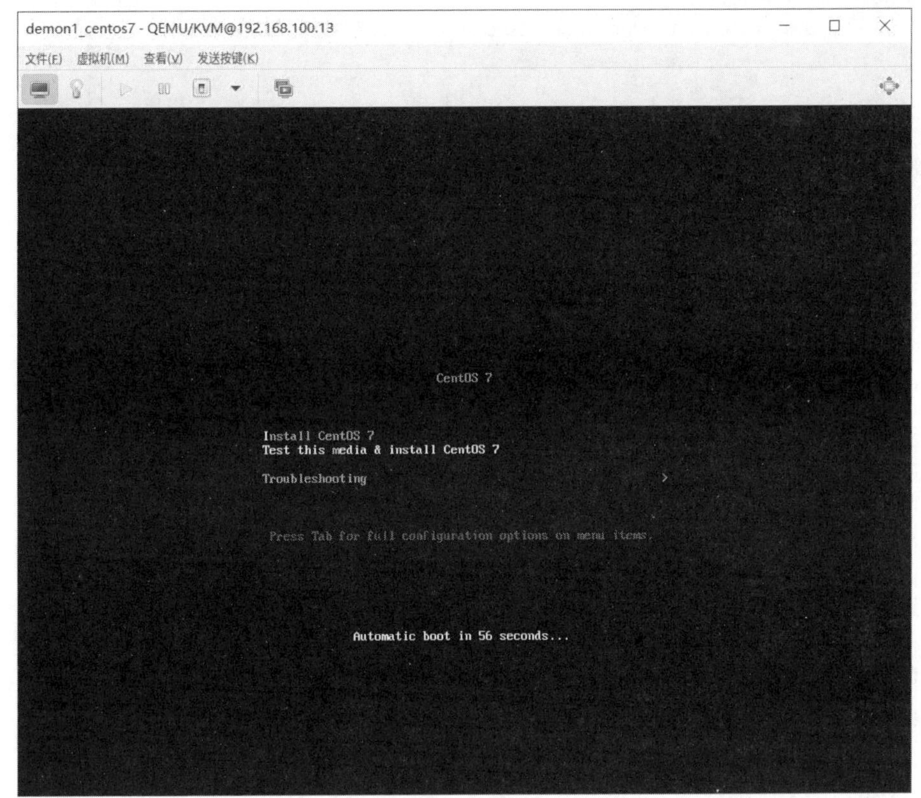

图 3-20　开始安装

3.1.2　使用 Virt-Manager 查看当前配置

上一章介绍了如何在图形化界面和系统最小化界面下启动 Virt-Manager，打开 Virt-Manager 主界面，最先看到的就是一个虚拟机的列表，如图 3-21 所示。

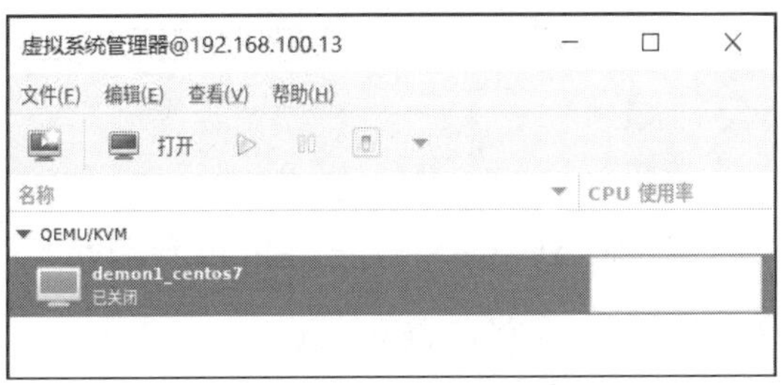

图 3-21　Virt-Manager 的主界面

这里根据虚拟的状态不同，双击虚拟机名称或者单击“打开”按钮可以看到虚拟机的显示信息。在虚拟机未打开的情况下可以通过单击□按钮来打开虚拟机，打开后可以看到如图 3-22 所示的页面。

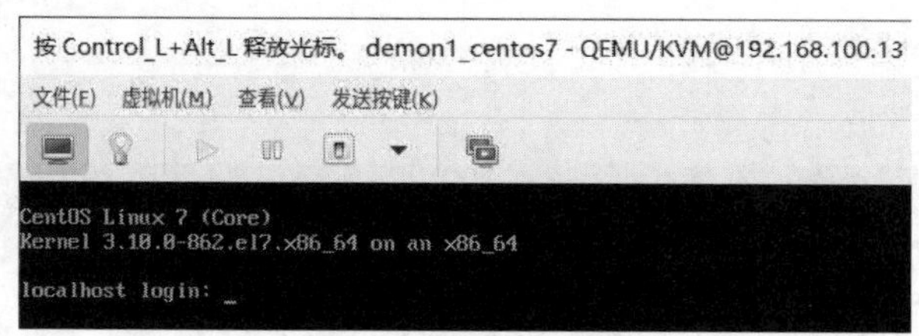

图 3-22　虚拟机控制台

单击工具栏中的图标，就会显示此虚拟机的详细配置，可以在这里查看或修改虚拟机的配置，如图 3-23 所示。

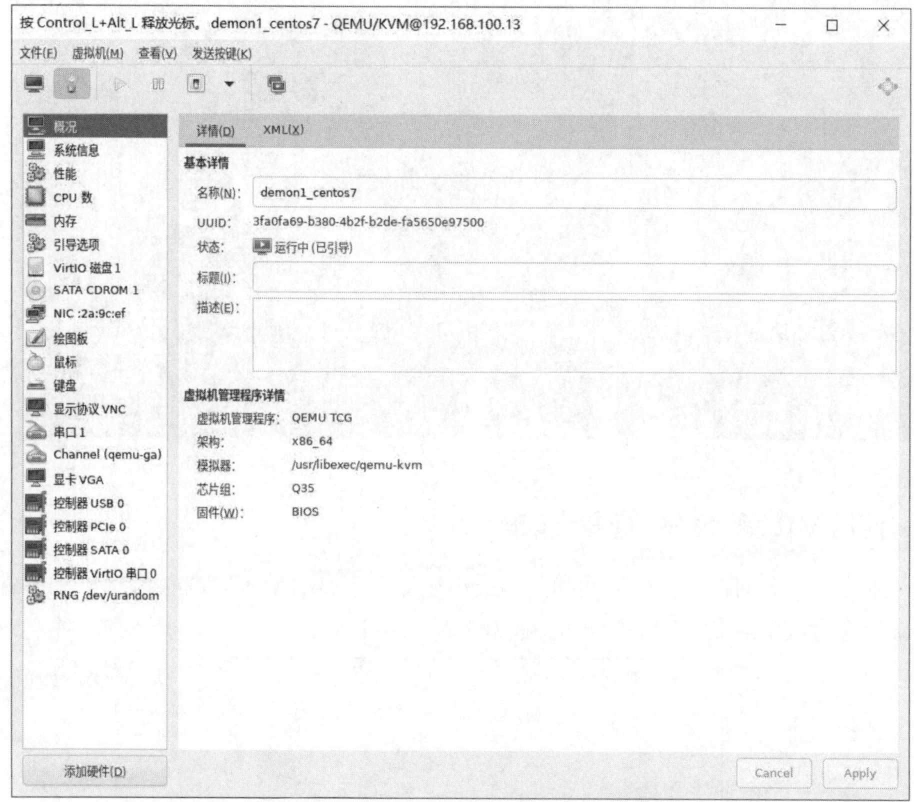

图 3-23　Virt-Manager 中的虚拟机配置管理

在 Virt-Manager 主界面中可以看到当前虚拟机的情况，也可以根据需求进行调整。

3.2 virt-install 命令创建虚拟机

与前面的创建方法相比，通过 virt-install 创建新虚拟机是效率最高的方法，同时也是最复杂的方法。

通过 virt-install 创建虚拟机需要满足两个先决条件：

1. 可访问的保存在本地网络上的操作系统安装源

可以是以下其中之一：

(1) 安装介质的 ISO 镜像文件。

(2) 现有操作系统的虚拟磁盘镜像。

2. 可实现的安装模式

(1) 如果是交互式安装，则需要 virt-viewer 软件。

(2) 如果是非交互式安装，就需要为操作系统安装程序提供回答文件，例如 kickstart 文件。

3.2.1 创建虚拟机并通过交互模式安装

若要使用 virt-install 进行虚拟机的创建，系统就必须要支持此命令，这里可以通过 virt-install --version 命令来验证系统是否已经安装好了命令，若系统提示“未找到命令”或“Command not found”，均表示系统当前没有 virt-install 命令，需要进行安装，我们可以通过以下命令来安装 virt-install。

```
# dnf install -y libguestfs-tools virt-viewer virt-install.noarch
```

在使用 virt-install 命令安装虚拟机的时候，必须要提供以下参数。

(1) --name：虚拟机的名称。

(2) --memory：虚拟机的内存。

(3) --vcpus：虚拟机虚拟 CPU(vCPU)的数量。

(4) --disk：虚拟机磁盘类型的大小。

(5) 操作系统安装源的类型和位置，可以由--location、--cdrom、--pxe、--import 和--boot 选项来指定。

可以通过查看 virt-install 的帮助信息获取帮助，命令如下：

```
# virt-install --help
```

若要查看 virt-install 选项属性的完整列表，命令如下：

```
# virt-install --option=?
```

当需要查看磁盘存储和选项的命令时，可以使用以下命令进行查看：

```
#virt-install --disk=?
```

当然，最完整的帮助信息在 virt-install 手册中，该手册中除了有对每个命令选项的介绍之外，还有重要的提示和丰富的示例。

运行 virt-install --version 和 virt-viewer -V 时出现以下信息即表示已经正常安装了，接下来就可以开始创建虚拟机，代码运行如下：

```
# virt-install --version
    4.1.0
# virt-viewer -V
    virt-viewer 的 11.0-1.el9 版本 (OS ID: rhel9)
```

下面创建一个名为 demo1_centos7 的虚拟机，给虚拟机分配 1024MB 内存，1 个 vCPU、8GB 的虚拟磁盘。将通过存储在本地/iso 目录中的 CentOS-7-x86_64-DVD-1804.iso 在虚拟机中安装操作系统。另外，通过--os-variant 指定虚拟机操作系统的版本为 CentOS 7。命令如下：

```
# virt-install \
--name centos7 \
--memory 1048 \
--vcpus 1 \
--disk size=8 \
--cdrom /iso/CentOS-7-x86_64-DVD-1804.iso    \
--os-variant centos7
    WARNING    KVM acceleration not available, using 'qemu'

    开始安装......
    正在分配 'centos7.qcow2'                    |     0 B    00:00:00 ...
    创建域......                                |     0 B    00:00:00
    正在运行图形控制台命令：virt-viewer --connect qemu:///system --wait centos7
```

由于仅指定虚拟磁盘的大小而未指定其存储位置和名称，所以 virt-install 会自动在默认的 default 存储池中创建一个名为 centos7.qcow2 的存储卷，然后分配给虚拟机。

当虚拟机创建完毕后，virt-install 会自动启动 virt-viewer，然后就可以在其中通过交互模式进行安装了，如图 3-24 所示。

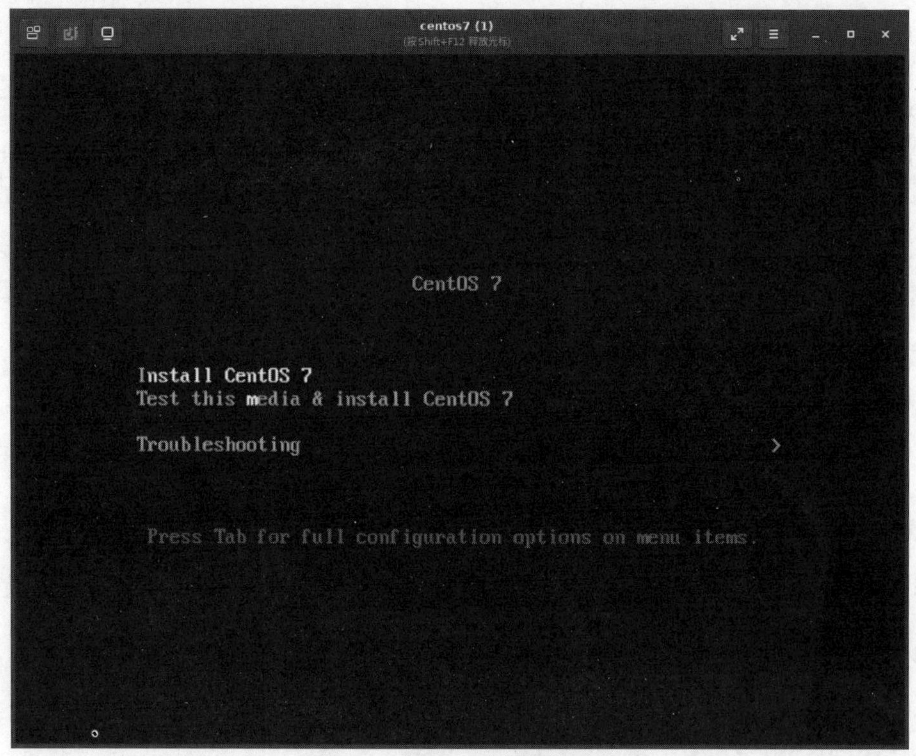

图 3-24　为 KVM 安装操作系统

3.2.2 查看虚拟机与环境的配置

安装完成之后，查看虚拟机的信息及宿主机存储的变化。

```
# virsh list
  Id    名称          状态
----------------------
  2     centos7     运行

# virsh dominfo centos7
Id:             2
名称:           centos7
UUID:           e3e073f1-5af1-4e22-a5f1-190ec00dcad0
OS 类型:        hvm
状态:           运行
CPU:            1
CPU 时间:       358.9s
最大内存:       1073152 KiB
使用的内存:     1073152 KiB
持久:           是
自动启动:       禁用
管理的保存:     否
安全性模式:     selinux
安全性 DOI:     0
安全性标签:     system_u:system_r:svirt_tcg_t:s0:c184,c501 (enforcing)
消息:           污染的：使用已弃用的配置设置
                弃用的配置：CPU 型号 'qemu64'
# virsh domblklist centos7
 目标     源
---------------------------------------------
 vda      /var/lib/libvirt/images/centos7.qcow2
 sda      -

# virsh vol-list default
 名称                        路径
------------------------------------------------------------------
 centos7.qcow2               /var/lib/libvirt/images/centos7.qcow2
 demon1_centos7.qcow2        /var/lib/libvirt/images/demon1_centos7.qcow2
```

3.2.3 virt-install 高级用法示例

virt-install 有很丰富的选项参数以适用不同的场景，下面来看示例。

示例：通过回答文件进行非交互式的自动化安装。

在手工安装 RHEL/CentOS 时，需要设置多个选项参数，很繁琐。可以使用 kickstart 文件实现自动化安装。kickstart 文件是一个包含安装程序所需要的选项参数的文本文件。

默认情况下 RHEL/CentOS 的安装程序会在/root 目录下生成一个名为 anaconda-ks.cfg 的文件，可以以这个文件的内容为“起点”快速生成一个回答文件，命令如下：

```
# cp anaconda-ks.cfg centos7.txt
```

```
# vi centos7.txt
    #可以根据需求进行修改。
    # Generated by Anaconda 34.25.2.1
    # Generated by pykickstart v3.32
    #version=RHEL9
    # Use graphical install
    graphical
    repo --name="AppStream" --baseurl=file:///run/install/sources/mount-0000-cdrom/AppStream

    %addon com_redhat_kdump --enable --reserve-mb='auto'

    %end

    # Keyboard layouts
    keyboard --xlayouts='cn'
    # System language
    lang zh_CN.UTF-8

    # Use CDROM installation media
    cdrom

    %packages
    @^minimal-environment

    %end

    # Run the Setup Agent on first boot
    firstboot --enable

    # Generated using Blivet version 3.6.0
    ignoredisk --only-use=nvme0n1
    autopart
    # Partition clearing information
    clearpart --none --initlabel

    # System timezone
    timezone Asia/Shanghai --utc

    # Root password
    rootpw --iscrypted --allow-ssh     $6$Jvg1S7iHrXsO2A/F$748GBJRtW4uooWfbHpKVBkSTwk1i2
    YNe73PO9JmUuNfuqDeCr7A2tJLbOkr     UkwhCCE0z6iyNJMCxCP7nYbB8m0
```

除了这个回答文件，还需要为 virt-install 命令提供 3 个选项参数。

(1) --location：为 virt-install 指定 Linux 发行版本的安装源。既可以是网络位置，也可以是本地 ISO 文件，例如/iso/CentOS-7-x86_64-DVD-1804.iso。

(2) --initrd-inject：需要与--location 选项一起使用，它指定一个保存在宿主机本地的 Kickstart 文件，例如/root/centos7.txt。

(3) --extra-args：需要与--location 选项一起使用，它传递给安装程序额外的参数，例如，通过“ks=file:/centos7.txt”指定 kickstart 文件。

示例命令如下：

```
# virt-install \
--name centos7 \
--memory 1048 \
--vcpus 1 \
--disk size=8 \
--os-variant centos7 \
--location /iso/CentOS-7-x86_64-DVD-1804.iso \
--initrd-inject /root/centos7.txt \
--extra-args="ks=file:/centos7.txt"
    WARNING    KVM acceleration not available, using 'qemu'

    开始安装......
    正在检索 ' vmlinuz'                    |    0 B    00:00:00 ...
    正在检索 ' initrd.img'                 |    0 B    00:00:00 ...
    正在分配 'centos7.qcow2'               |    0 B    00:00:00 ...
    创建域......                            |    0 B    00:00:00
    正在运行图形控制台命令：virt-viewer --connect qemu:///system --wait centos7
```

虚拟机创建完毕之后，上述这个 virt-install 命令并不会启动 virt-viewer。只要 kickstart 文件内容正确，就会在非交互模式下进行自动安装。当然，也可以通过 Cockpit 或 virt-manager 中的控制台查看安装过程。

3.3 VNC 连接 KVM 虚拟机

前面讲解了如何创建 KVM 虚拟机，KVM 虚拟机和 VMware 虚拟机一样，有时候只启动就行，其他工作都需要在 XShell 或 Putty 等远程连接软件上完成，我们都知道 Windows 里面有远程桌面可以直接实现远程连接，而在 KVM 中，也有类似的软件可以直接连接 KVM 虚拟机，即本节要介绍的 VNC 连接 KVM 虚拟机。

3.3.1 什么是 VNC

所谓 VNC(virtual network console)，就是虚拟网络控制台的缩写，它是一款优秀的远程控制工具软件，由 AT&A 欧洲的研究实验室开发。VNC 是基于 UNIX 和 Linux 操作系统的免费的开源软件，远程控制能力强大，高效实用，性能足够媲美 Windows 和 MAC 中的任何远程控制软件。

VNC 基本由两部分组成，一部分是客户端的应用程序，另外一部分是服务器端的应用程序。VNC 的基本运行原理和一些 Windows 的远程控制软件很像。VNC 的服务端应用程序在 UNIX 和 Linux 操作系统中适应性很强，图形用户界面十分友好，看上去和 Windows 下的软件界面也很类似。

要连接虚拟机，还可以通过 XManager，这里还要介绍 VNC，主要是因为 VNC 和 XManager 的工作原理不一样。VNC 是远程连入操作系统，所有操作在 UNIX、Linux 主机服务端进行，即使操作过程中“本地电脑与操作主机网络断开”，也不影响操作的顺利进行，而 XManager 是通过端口将主机服务器的 UI 界面引导到本地电脑进行展现，如操作过程中出现“本地电脑与操作主机网络断开”，操作将中断、失败。更重要的是 VNC 是免费的、开源的。

3.3.2 VNC 服务端

所谓 VNC 服务端，就是指为客户端提供服务的应用程序，一般运行在服务器上，那么服务端要怎样安装呢？

首先，需要先安装 tigervnc 的服务端，安装命令如下：

```
# dnf install -y tigervnc-server
```

接下来，需要对 VNC Server 进行配置。

首先，需要对 vncserver@.service 进行编辑，这里由于涉及关键的文件，需要先对源文件进行备份，以便以后需要恢复的时候使用。

注意

对于操作系统中的重要文件，建议先对源文件进行备份处理，以便日后需要对文件进行恢复时便于操作。

```
# cp /usr/lib/systemd/system/vncserver@.service \ /etc/system/system/vncserver@.service
# vi /etc/system/system/vncserver@.service
     [Unit]
    Description=Remote desktop service (VNC)
    After=syslog.target network.target

     [Service]
    Type=simple

    # Clean any existing files in /tmp/.X11-unix environment
    ExecStartPre=/bin/sh -c '/usr/bin/vncserver -kill %i > /dev/null 2>&1 || :'
    ExecStart=/usr/bin/vncserver_wrapper root %i
    ExecStop=/bin/sh -c '/usr/bin/vncserver -kill %i > /dev/null 2>&1 || :'

     [Install]
    WantedBy=multi-user.target
```

其次，需要重新载入以下命令。

```
#systemctrl daemon-reload
```

然后，设置 VNC 的密码。

```
# vncpasswd
    Password:
```

```
Verify:
```

紧接着，启动 VNC Server 服务。

```
# systemctl start vncserver@:1.service
```

我们可以根据自己的需求设置是否需要开机启动服务，如果需要则可以执行下面的代码：

```
# systemctl enable vncserver@:1.service
```

到这里服务端就设置好了，接下来可以通过下面的命令查看服务的状态。

```
# systemctl status vncserver@:1.service
```

至此服务端就完全设置好了，接下来是客户端，客户端设置好之后就可以进行远程连接了。

3.3.3 VNC 客户端

既然 VNC 有服务端，那么肯定就需要客户端。这里以 REALVNC 为例，REALVNC 的下载地址为 https://www.realvnc.com/en/connect/download/viewer/，这是官方的下载地址，因为软件本身就是免费的，所以建议大家直接去官方网站下载，如图 3-25 所示。

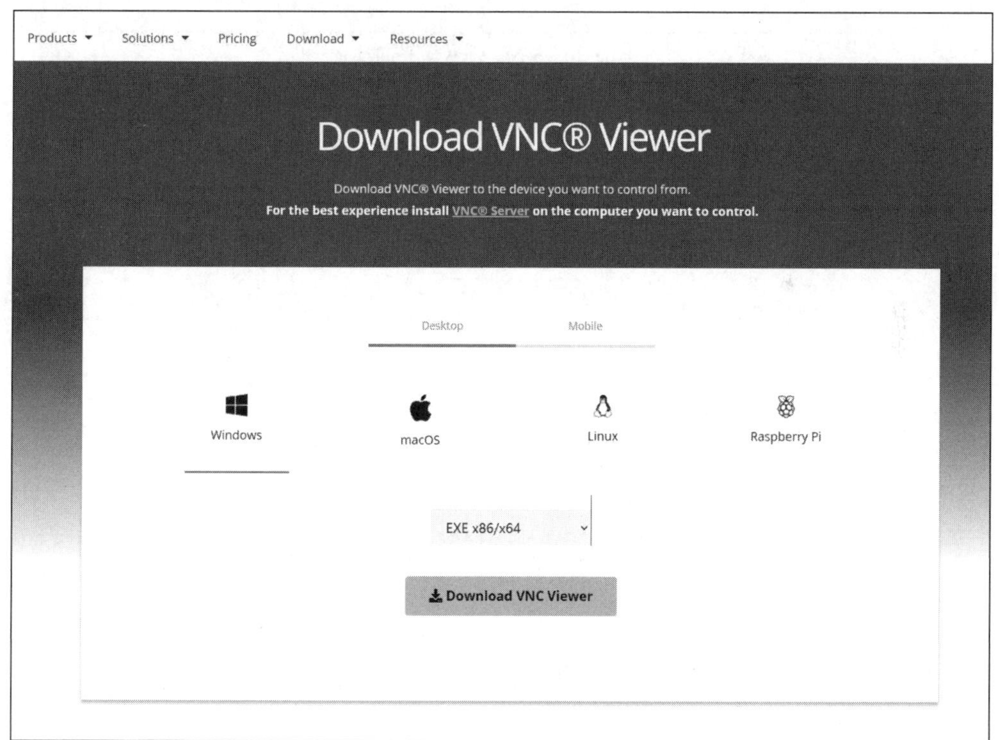

图 3-25　REALVNC 官网

下载安装后就可以在“开始”菜单中看到刚安装的 VNC 连接软件，如图 3-26 所示。

图 3-26　VNC Viewer 界面

打开软件后就可以看到软件的主界面，也可以通过 File→New Connection 创建与服务器的连接，VNC Server 中的 home 对应我们的服务器 IP，如图 3-27 所示。

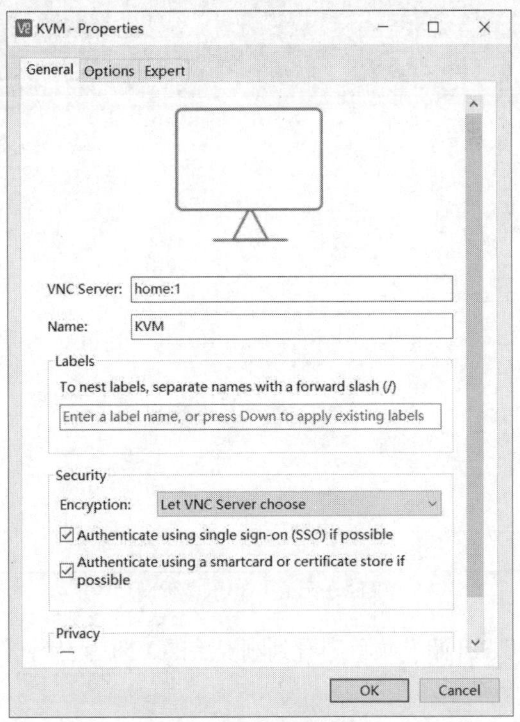

图 3-27　创建 VNC 连接

注意

这里需要注意的是主机必须要 ping 通服务器，才可以进行连接。

创建好之后 VNC 主界面就会出现一个 KVM 图标，双击该图标就可以进行连接(图 3-28)，在验证完密码之后就可以正常使用 VNC 进行远程连接了。

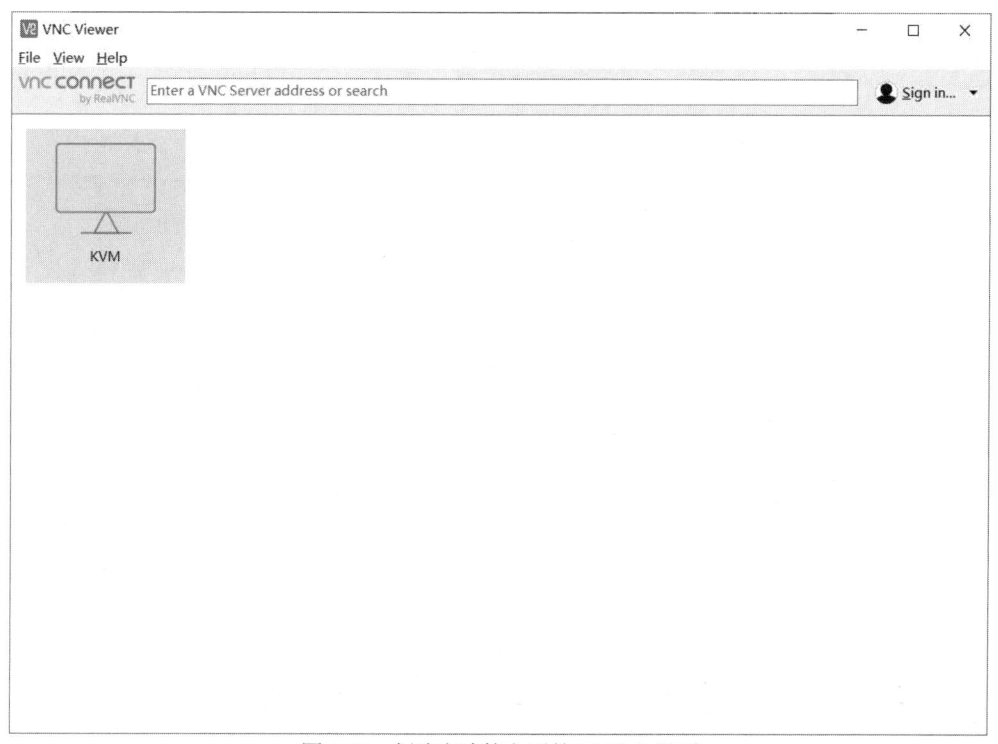

图 3-28　创建完连接之后的 VNC 主界面

3.4 习题

1. 简述如何使用 Virt-Manager 图形化创建 KVM 虚拟机。
2. 简述如何使用 virt-install 命令创建虚拟机。
3. 简述如何使用 VNC 连接到虚拟机。

第4章

虚拟机管理

在RHEL/CentOS系统中，最容易掌握的虚拟机管理工具是图形化的Cockpit和Virt-Manager。但在大多数情况下，服务器的操作系统基本都是最小化安装(minimal install)。所以我们使用最多的是效率最高、功能最强大的命令行管理工具virsh。

本章要点

- ◎ libvirt架构概述
- ◎ 使用virsh管理虚拟机

4.1 libvirt架构描述

在KVM中，libvirt是一个管理虚拟机及其他虚拟化功能的软件合集，包括API库、守护进程(libvirtd)和管理工具(virsh 工具集)。KVM将libvirt作为虚拟化管理的引擎，而Cockpit、Virt-Manager及virsh等管理工具都是通过libvirtd进行管理操作。它们将请求发给libvirtd，libvirtd根据配置文件对虚拟化平台的技术、存储、网络等资源进行管理，如图4-1所示。

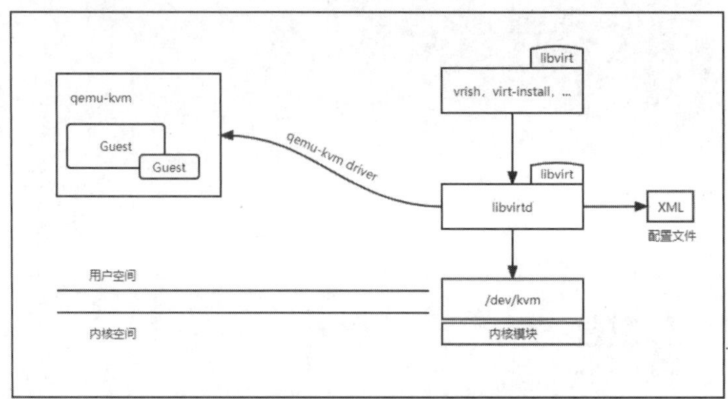

图 4-1　libvirt 是 KVM 虚拟化管理的引擎

在 RHEL/CentOS 中，要正常管理虚拟化平台，就一定要保证 libvirtd 守护程序的正常运行。下面通过实验来验证：

```
# virsh list --all
 Id     Name                  State
--------------------------------------------
 -      demo1_centos7      shut off

# systemctl stop libvirtd

# virsh list --all
 Id     Name                  State
-----------------------------------------
 -      demo1_centos7      shut off

# virsh list --all
error: failed to connect to the hypervisor
error: Cannot recv data: Connection reset by peer

# systemctl start libvirtd

# virsh list --all
 Id     Name                  State
-----------------------------------------
 -      demo1_centos7      shut off
```

从上述这个实验可以看出：如果 libvirtd 处于停止状态，则连最基本的 virsh list 命令都会执行失败。

4.2 使用 virsh 管理虚拟机

virsh 是 Virtual Shell 的缩写，它采用命令行用户界面管理虚拟化平台。我们可以使用 virsh 来创建、列出、编辑、启动、重新启动、停止、挂起、恢复、关闭和删除虚拟机等。virsh 命令的功能十分强大，甚至一些复杂的操作也可以完成，除了 KVM 之外，它还支持 LXC、Xen、QEMU、OpenVZ、Virtualbox 和 VMware ESX/ESXi。

本章先讲解 virsh 最常用的一些功能。

4.2.1 获得帮助

除了可以使用 man virsh 查看帮助手册之外，还可以通过它的 help 选项查看可用命令列表及简要说明，命令如下：

```
# virsh --help

virsh [options]... [<command_string>]
virsh [options]... <command> [args...]

  options:
    -c | --connect=URI          hypervisor connection URI
    -d | --debug=NUM            debug level [0-4]
    -e | --escape <char>        set escape sequence for console
    -h | --help                 this help
    -k | --keepalive-interval=NUM keepalive interval in seconds, 0 for disable
    -K | --keepalive-count=NUM number of possible missed keepalive messages
    -l | --log=FILE             output logging to file
    -q | --quiet                quiet mode
    -r | --readonly             connect readonly
    -t | --timing               print timing information
    -v                          short version
    -V                          long version
    --version[=TYPE]     version, TYPE is short or long (default short)
  commands (non interactive mode):

 Domain Management (help keyword 'domain')
    attach-device               attach device from an XML file
    attach-disk                 attach disk device
    attach-interface            attach network interface
    autostart                   autostart a domain
    blkdeviotune                set or query a block device I/O tuning parameters
    blkiotune                   get or set blkio parameters
    blockcommit                 start a block commit operation
    blockcopy                   start a block copy operation
    blockjob                    manage active block operations
……
```

当第一眼看到 virsh 命令的帮助信息时，你可能会觉得它的内容太多，但也能从侧面感受到它的功能强大。当然我们不必记住所有内容，只阅读要运行并使用的子命令的说明即可。这些子命令分为以下几类：

(1) 域管理(domain management)。

(2) 域监控(domain monitoring)。

(3) 宿主机和管理程序(host and hypervisor)。

(4) 接口(interface)。
(5) 网络过滤器(network filter)。
(6) 网络(networking)。
(7) 节点设备(node device)。
(8) 密语(secret)。
(9) 快照(snapshot)。
(10) 存储池(storage pool)。
(11) 存储卷(storage volume)。
(12) virsh 自身的子命令。

提示

在 KVM 中，域(domain)与虚拟机(virtual machine, vm)、快照(snapshot)和检查点(checkpoint)是同义词，可以进行互换。

每类子命令都包含与执行特定任务相关的命令。我们可以单独查看某类子命令的帮助，例如，查看与虚拟机(域)管理相关的子命令的帮助，命令如下：

```
# virsh help domain
 Domain Management (help keyword 'domain'):
    attach-device                  attach device from an XML file
    attach-disk                    attach disk device
    attach-interface               attach network interface
    autostart                      autostart a domain
    blkdeviotune                   set or query a block device I/O tuning parameters
    blkiotune                      get or set blkio parameters
    blockcommit                    start a block commit operation
    blockcopy                      start a block copy operation
    blockjob                       manage active block operations
    blockpull                      populate a disk from its backing image
    blockresize                    resize block device of domain
    … …
```

还可以进一步显示特定子命令的帮助。例如，查看 list 子命令的帮助，命令如下：

```
# virsh help list
  NAME
    list - list domains

  SYNOPSIS
    list [--inactive] [--all] [--transient] [--persistent] [--with-snapshot]
      --without-snapshot][--with-checkpoint][--without-checkpoint] [--state-running] [--state-paused]
      [--state-shutoff] [--state-other] [--autostart] [--no-autostart] [--with-managed-save] [--without-managed-save]
      [--uuid] [--name] [--id] [--table] [--managed-save] [--title]

  DESCRIPTION
    Returns list of domains.
```

```
OPTIONS
  --inactive          list inactive domains
  --all               list inactive & active domains
  --transient         list transient domains
```

4.2.2 常用子命令

前面介绍 virsh 命令合集的时候，估计有很多人都会被它的内容所“震撼”，我们没必要记住所有命令，下面列出经常用到的一些子命令。

1. 列出虚拟机

获得处于运行或挂起(suspend)模式的虚拟机的列表，命令如下：

```
# virsh list
 Id    Name                State
----------------------------------------------
 1     demo1_centos7       running
```

可以使用--inactive 选项只显示不活动的虚拟机。

如果想查看所有状态的虚拟机，则应执行的命令如下：

```
# virsh list --all
 Id    Name                State
----------------------------------------------
 1     demo1_centos7       running
 -     demo2_centos7       shut off
 -     demo3_centos7       shut off
```

从以上信息中可以看到：有 1 台虚拟机(域)处于 running 状态，libvirt 给它分配的 ID 号是 1。其他虚拟机的状态是 shut off，而且没有 ID 号。

2. 启动虚拟机

启动虚拟机，就是打开、运行虚拟机，例如，要打开名称为 demo1_centos7 的虚拟机，命令如下：

```
# virsh start demo1_centos7
 Domain demo1_centos7 started
```

提示

virsh 像大多数 Linux 命令一样，具有自动完成功能。可以使用键盘上的 Tab 键完成子命令、选项和参数的自动补全。

验证虚拟机是否正在运行，命令如下：

```
# virsh list
 Id    Name                State
----------------------------------------------
```

```
        1      demo1_centos7     running
        2      demo2_centos7     running
        -      demo3_centos7     shut off
```

3. 关闭虚拟机

使用 shutdown 子命令可正常关闭虚拟机。在使用时，可以通过名称也可以通过 ID 指定虚拟机，命令如下：

```
# virsh list
     Id     Name                    State
    --------------------------------------------
      1      demo1_centos7     running

# virsh shutdown 1
Domain '1' is being shutdown

# virsh list
     Id     Name                State
    --------------------------------------
    1      demo1_centos7     running
    2      demo2_centos7     running
```

由于 shutdown 子命令需要与虚拟机的操作系统一起协调工作，所以不能保证它会立即成功地关闭虚拟机。关闭所需时间的长短，也取决于虚拟机中要关闭的服务和进程的多少。

如果 shutdown 子命令无法关闭虚拟机，则只能通过 destroy 子命令强制关闭虚拟机了，这与把电源线直接从物理机上拔出很类似，命令如下：

```
# virsh destroy demo1_centos7
     Domain 'demo1_centos7' destroyed
```

提示

每台运行着的虚拟机都是宿主机上的一个进程(/usr/libexec/qemu-kvm)，但是不推荐使用 kill 命令结束进程的方法来关闭虚拟机。

4. 重启虚拟机

使用 reboot 子命令可以重新启动虚拟机，就像从控制台中运行重新启动命令一样。这个子命令当然也需要与虚拟机的操作系统协同工作，命令如下：

```
# virsh reboot demo1_centos7
     Domain 'demo1_centos7' is being rebooted
```

如果 reboot 子命令无法重新启动虚拟机，则只能通过 reset 子命令来强制重新启动虚拟机，这与按下物理机上的重置按钮类似。命令如下：

```
# virsh reset demo1_centos7
     Domain 'demo1_centos7' was reset
```

注意

destroy 和 reset 子命令有可能会造成虚拟机数据的损坏，需谨慎使用。

5. 挂起(暂停)与恢复虚拟机

使用 suspend 子命令可以挂起(暂停)正在运行的虚拟机，而使用 resume 子命令又可以将它们恢复运行。命令如下：

```
# virsh list
   Id    Name               State
  ----------------------------------------
   1     demo1_centos7      running

# virsh suspend demo1_centos7
Domain 'demo1_centos7' suspended

# virsh list
   Id    Name               State
  ----------------------------------------
   1     demo1_centos7      paused

# virsh resume demo1_centos7
Domain 'demo1_centos7' resumed

# virsh list
   Id    Name               State
  -----------------------------------------
   1     demo1_centos7      running
```

6. 保存与还原虚拟机

可以将正在运行的虚拟机的内存数据保存到状态文件(state file)中，这类似于在虚拟机操作系统中进行休眠操作。保存虚拟机之后，该虚拟机将不再在宿主机上运行，因此会释放原来分配给该虚拟机的资源。保存虚拟机的命令如下：

```
# virsh save demo1_centos7 demo1_centos7-save
  Domain 'demo1_centos7' saved to demo1_centos7-save

# ls -lh demo1_centos7-save
  -rw-------. 1 root root 5.9M Nov1    15:40 demo1_centos7-save

# file demo1_centos7-save
  demo1_centos7-save: Libvirt QEMU Suspend Image, version 2, XML length 73314, running

# virsh list --all
  Id    Name               State
 ------------------------------------------
  -     demo1_centos7      shut off
  -     demo2_centos7      shut off
```

```
    -      demo3_centos7      shut off
```

保存虚拟机需要一些时间，这取决于虚拟机内存数据的多少。

当需要虚拟机的时候，可以进行还原操作。例如，从状态文件 demo1_centos7-save 还原虚拟机的命令如下：

```
# virsh restore demo1_centos7-save
  Domain restored from demo1_centos7-save

# virsh list
    Id    Name                  State
  ----------------------------------------------
    1     demo1_centos7     running
```

7. 查看、编辑虚拟机配置文件

默认情况下，libvirtd 使用保存在/etc/libvrit/qmeu/目录中的虚拟机配置文件，命令如下：

```
# ls -l /etc/libvirt/qemu
    total 24
    drwx------. 2 root root 6 Aug         5 00:30    autostart
    -rw-------. 1 root root 7876 Oct 24   11:02      demo1_centos7.xml
    -rw-------. 1 root root 7876 Nov 1    15:17      demo2_centos7.xml
    -rw-------. 1 root root 7876 Nov 1    15:19      demo3_centos7.xml
    drwx------. 3 root root 42 Sep 26     11:03      networks
```

一个虚拟机对应一个 XML 格式的配置文件。除了可以使用 cat 等命令查看配置文件之外，还可以使用 virsh 的 dumpxml 子命令查看虚拟机的当前配置，命令如下：

```
# virsh dumpxml demo1_centos7
  <domain type='kvm' id='1'>
    <name>demo1_centos7</name>
    <uuid>6fbb4b60-0e05-4d1e-8eab-269d0b0566b6</uuid>
    <metadata xmlns:libosinfo="http://libosinfo.org/xmlns/libvirt/domain/1.0"
      xmlns:cockpit_machines="https://github.com/cockpit-project/cockpit-machines">
      <libosinfo:libosinfo>
        <libosinfo:os id="http://centos.org/centos/7.0"/>
      </libosinfo:libosinfo>
      <cockpit_machines:data>
        <cockpit_machines:has_install_phase>false</cockpit_machines:has_install_phase>
        <cockpit_machines:install_source_type>file</cockpit_machines:install_source_type>
        <cockpit_machines:install_source>/iso/CentOS-7-x86_64-DVD-1804.iso</cockpit_
              machines:install_source>
  ……
```

注意

有关配置文件中元素和属性的详细解释可参阅 libvirt 的官方文档：https://libvirt.org/formatdomain.html。

我们可以通过编辑配置文件来修改虚拟机的设置，这种修改在虚拟机下次启动时才会生效。不建议通过 vi 等文本编辑软件直接修改配置文件，而是使用 virsh 的子命令 edit。edit 子命令具有一些错误检查的功能，命令如下：

```
# virsh edit demo1_centos7
```

提示

edit 子命令将使用由$EDITOR 变量设置的编辑器打开文件，默认为 vi 编辑器。

edit 子命令会自动将配置文件复制为一个新的临时文件，然后调用文件编辑器打开这个文件。当修改保存时，会先进行格式检查，没有问题之后才会再覆盖原有的配置文件。

8. 创建新的虚拟机

除了使用 virt-install、virt-manager、Cockpit 等工具创建新的虚拟机之外，还可以使用 virsh 的 create 或 define 子命令创建新的虚拟机。

create 子命令可以创建新的临时性虚拟机，如果同时指定--paused 选项，则新虚拟机的状态为暂停状态，否则将处于运行状态。由于 create 子命令创建的是临时虚拟机，所以不会在/etc/libvirt/qemu 目录中创建新虚拟机的配置文件。一旦关闭临时虚拟机，libvirt 就不知道它的存在了。

与 create 命令正好相反，define 子命令会在/etc/libvirt/qemu/目录中创建新虚拟机的配置文件，所以是永久性虚拟机，但是由于新虚拟机处于关闭状态，还需要使用 start 子命令来启动它。

create 和 define 子命令都需要使用 XML 格式的配置文件。我们既可以根据 libvirt 的 XML 格式要求(https://libvirt.org/format.html)来全新创建，也可以参考现有的虚拟机的配置文件来创建。最简单的方法是使用 dumpxml 从现有的虚拟机中先复制一个基础配置，然后进行修改。现在，做一个实验，生成一个新的虚拟机：demo5_centos7，示例如下：

```
1 # virsh dumpxml demo1_centos7 > new5.xml

2 # virsh domblklist demo1_centos7
 Target     Source
------------------------------------------------------
 vda        /var/lib/libvirt/images/demo1_centos7.qcow2
 sda        -

3 # cd /var/lib/libvirt/images/

4 # cp -p demo1_centos7.qcow2 demo5_centos7.qcow2

5 # vi /root/new5.xml
```

注意

在使用 virtsh dumpxml 命令导出 XML 配置文件时，注意将相应虚拟机域关机之后，再进行导出操作，这样可以省略以后很多复杂的操作。

为了避免两台虚拟机使用相同的虚拟机磁盘文件，所以通过第 2 行命令获得原来虚拟机的磁盘文件的信息，第 3、4 行命令复制出一个新的磁盘文件供新的虚拟机使用。

在编辑 XML 文件时，要注意新虚拟机的名称要保持唯一，将<<UUID>>、<<mac address>>删除，同时要使用新的虚拟磁盘文件，代码如下：

```
<domain type='kvm'>
  <name>demo5_centos7</name>
  <uuid>6fbb4b60-0e05-4d1e-8eab-269d0b0566b6</uuid>
……
    <disk type='file' device='disk'>
      <driver name='qemu' type='qcow2' discard='unmap'/>
      <source file='/var/lib/libvirt/images/demo5_centos7.qcow2'/>
      <target dev='vda' bus='virtio'/>
      <address type='pci' domain='0x0000' bus='0x04' slot='0x00' function='0x0'/>
    </disk>
    <disk type='volume' device='disk'>
……
    <interface type='network'>
      <mac address='52:54:00:8b:f7:e3'/>
      <source network='default'/>
      <model type='virtio'/>
      <address type='pci' domain='0x0000' bus='0x01' slot='0x00' function='0x0'/>
    </interface>
……
```

提示

新生产 new5.xml 配置文件中，一共做了 4 处修改，分别如下：

(1) 新虚拟机的名称。

(2) UUID。

(3) mac address。

(4) disk。

有了这个新的 XML 文件，下面就可以创建新的虚拟机。先尝试用 create 子命令创建新的虚拟机，命令如下：

```
6 # virsh create new5.xml
  Domain 'demo5_centos7' created from new5.xml

7 # virsh list
    Id     Name                State
```

```
    -------------------------------------------
     1      demo5_centos7     running

8 # ls /etc/libvirt/qemu
  autostart demo1_centos7.xml   demo2_centos7.xml   demo3_centos7.xml   networks

9 # virsh destroy demo5_centos7
  Domain 'demo5_centos7' destroyed

10 # virsh list --all
     Id    Name                  State
    -----------------------------------------
     -     demo1_centos7         shut off
     -     demo2_centos7         shut off
     -     demo3_centos7         shut off
```

从第 8 行命令的输出可以看出，create 子命令并不会创建新虚拟机的配置文件，所以如果将虚拟机关闭，libvirt 就不会知道有过这个虚拟机，这可以从第 10 行命令的输出中得到验证。

接下来使用 define 子命令创建新的虚拟机，命令如下：

```
11 # virsh define new5.xml
  Domain 'demo5_centos7' defined from new5.xml

12 # virsh list --all
     Id    Name                  State
    ------------------------------------------
     -     demo1_centos7         shut off
     -     demo2_centos7         shut off
     -     demo3_centos7         shut off
     -     demo5_centos7         shut off

13 # ls -lt /etc/libvirt/qemu
total 32
    -rw-------. 1 root root 7876   Nov 1    17:39    demo5_centos7.xml
    -rw-------. 1 root root 7876   Nov 1    15:19    demo3_centos7.xml
    -rw-------. 1 root root 7876   Nov 1    15:17    demo2_centos7.xml
    -rw-------. 1 root root 7876   Oct 24   11:02    demo1_centos7.xml
    drwx------. 3 root root 42     Sep 26   11:03    networks
    drwx------. 2 root root 6      Aug 5    00:30    autostart

14 # virsh start demo5_centos7
  Domain 'demo5_centos7' started
```

从第 12 行和第 13 行命令的输出可以看出：define 子命令创建的新虚拟机处于关闭状态，但是在/etc/libvirt/qemu/目录中会有相关配置文件。

9. 删除虚拟机

如果不再需要某个虚拟机，则可以先将其关闭，然后使用 undefine 子命令将配置文件删除。如果虚拟机正在运行，则 undefine 子命令不会停止它。默认情况下，undefine 子命令不会删除虚拟机的磁盘文件，命令如下：

```
1 # virsh list
   Id    Name                 State
 ------------------------------------------------
   1     demo5_centos7       running

2 # virsh destroy demo5_centos7
  Domain 'demo5_centos7' destroyed

3 # virsh undefine demo5_centos7
  Domain 'demo5_centos7' has been undefined

4 # virsh list --all |grep demo5

5 # rm /var/lib/libvirt/images/demo5_centos7.qcow2
  rm: remove regular file '/var/lib/libvirt/images/demo5_centos7.qcow2'? y
```

提示

如果虚拟机的磁盘文件是通过 libvirt 的存储池、存储卷进行管理的，则还可以使用 undefine 子命令中的--remove-all-storage、--storage 选项来同时清除它们。

4.3 习题

1. 创建一个全新虚拟机，名称为 new_vm。

2. 使用 virsh 管理此虚拟机，包括创建、列出、编辑、启动、重新启动、停止、挂起、恢复、关闭和删除虚拟机等操作。

第5章

管理KVM虚拟网络

本章将系统讲解虚拟网络的日常管理，这需要具备一定的 Linux 网络管理的知识，包括物理网络接口、虚拟网络接口、网桥及路由等。

本章要点

◎ TUN/TAP 设备工作原理与管理

◎ 网桥工作原理与管理

◎ 不同网络类型的原理

◎ NAT、桥接、隔离、路由、开发等网络类型的配置

◎ VLAN 的原理与配置

◎ 网络过滤器的原理与配置

5.1 查看默认网络环境

KVM 支持多种网络类型，首先查看一下默认的网络环境。

在 RHEL/CentOS 9 等 Linux 发行版中安装虚拟化组件的时候，通常会自动创建一个默认的虚拟网络配置，这包括一个命令 virbr0 的网桥、virbr0-nic 的虚拟网络接口、iptables 的 NAT 配置及 DNSMASQ 的配置等。下面就通过实验来查看默认的网络环境，从而理解 libvirt 虚拟网络的原理。

5.1.1 查看宿主机的网络环境

在没有启动虚拟机的情况下，宿主机上默认的网络环境如图 5-1 所示。

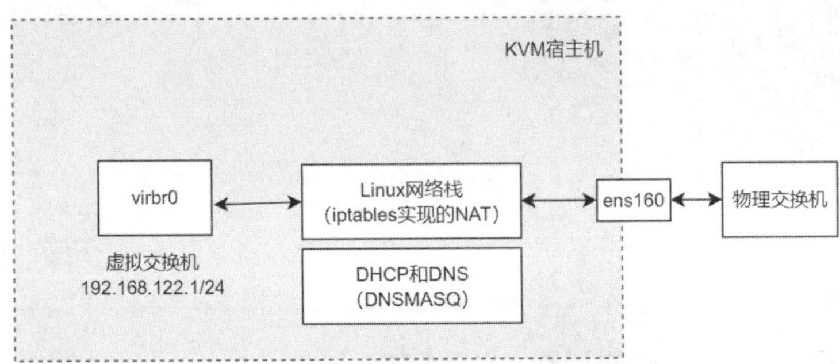

图 5-1 宿主机上默认的网络环境

查看宿主机网络环境的命令如下：

```
1 # ip address
1: lo: <LOOPBACK,UP,LOWER_UP> mtu 65536 qdisc noqueue state UNKNOWN group default qlen 1000
    link/loopback 00:00:00:00:00:00 brd 00:00:00:00:00:00
    inet 127.0.0.1/8 scope host lo
       valid_lft forever preferred_lft forever
    inet6 ::1/128 scope host
       valid_lft forever preferred_lft forever
2: ens160: <BROADCAST,MULTICAST,UP,LOWER_UP> mtu 1500 qdisc mq state UP group default qlen 1000
    link/ether 00:0c:29:5d:b1:9a brd ff:ff:ff:ff:ff:ff
    altname enp3s0
    inet 192.168.1.128/24 brd 192.168.1.255 scope global noprefixroute ens160
       valid_lft forever preferred_lft forever
    inet6 fe80::20c:29ff:fe5d:b19a/64 scope link noprefixroute
       valid_lft forever preferred_lft forever
3: virbr0: <NO-CARRIER,BROADCAST,MULTICAST,UP> mtu 1500 qdisc noqueue state DOWN group
    default qlen 1000
    link/ether 52:54:00:0f:b3:2b brd ff:ff:ff:ff:ff:ff
    inet 192.168.122.1/24 brd 192.168.122.255 scope global virbr0
       valid_lft forever preferred_lft forever

2 # ip link
1: lo: <LOOPBACK,UP,LOWER_UP> mtu 65536 qdisc noqueue state UNKNOWN mode DEFAULT group
    default qlen 1000
    link/loopback 00:00:00:00:00:00 brd 00:00:00:00:00:00
2: ens160: <BROADCAST,MULTICAST,UP,LOWER_UP> mtu 1500 qdisc mq state UP mode DEFAULT
    group default qlen 1000
    link/ether 00:0c:29:5d:b1:9a brd ff:ff:ff:ff:ff:ff
    altname enp3s0
3: virbr0: <NO-CARRIER,BROADCAST,MULTICAST,UP> mtu 1500 qdisc noqueue state DOWN mode
    DEFAULT group default qlen 1000
    link/ether 52:54:00:0f:b3:2b brd ff:ff:ff:ff:ff:ff

3 # nmcli connection
NAME    UUID                                  TYPE      DEVICE
ens160  60edd0a5-1c19-3b74-bfd5-8909037aec33  ethernet  ens160
virbr0  d5e1fa5c-cc7b-423f-a785-01e4fd87d8b0  bridge    virbr0
```

```
4 # nmcli device
DEVICE      TYPE        STATE                    CONNECTION
ens160      ethernet    connected                ens160
virbr0      bridge      connected (externally)   virbr0
lo          loopback    unmanaged                --
```

libvirt 网络的最重要组件是虚拟网络交换机，默认由 Linux 的网桥实现。libvirt 默认会创建一个名为 virbr0 的网桥，virbr0 是 Virtual Bridge 0 的缩写，与物理交换机类似，虚拟网络交换机也是从其接收的数据包(帧)中获得 MAC 地址，并存储在 MAC 表中。物理交换机的端口数量有限，而虚拟机交换机的端口数量则没有限制。

在 Linux 系统中，可以向网桥分配 IP 地址。从第 1 行命令的输出中可以看到，libvirt 向 virbr0 分配的 IP 地址是 192.168.122.1/24。libvirt 根据自己的配置文件分配这个 IP 地址，所以不会在 /etc/sysconfig/network-scripts/目录中看到名为 ifcfg-virbr0 的配置文件。

5.1.2 查看 libvirt 的网络环境

在 libvirtd 守护程序启动时，会根据配置文件创建一个名为 default 的虚拟网络。除了在宿主机上创建 virbr0 之外，还会通过配置 IP 转发、iptables 的 NAT 表，从而在 Linux 协议栈中实现 NAT 功能。

libvirt 使用的是 IP 伪装(IP masquerading)，而不是源 NAT(source NAT，SNAT)或目标 NAT(destination NAT，DNAT)，IP 伪装使虚拟机可以使用宿主机的 IP 地址与外部网络进行通信。默认情况下，当虚拟网络交换机以 NAT 模式运行时，虚拟机可以访问位于宿主物理计算机外部的资源，但是位于宿主物理计算机外部的计算机无法与内部的虚拟机进行通信，简单来说：仅允许由内到外，而不允许从外到内的访问。

下面通过命令行工具查看网络环境，命令如下：

```
1 # cat /proc/sys/net/ipv4/ip_forward
1

2 # iptables -t nat -L -n
  ……

Chain LIBVIRT_PRT (1 references)
target        prot  opt source            destination
RETURN        all   --  192.168.122.0/24  224.0.0.0/24
RETURN        all   --  192.168.122.0/24  255.255.255.255
MASQUERADE    tcp   --  192.168.122.0/24  !192.168.122.0/24   masq ports: 1024-65535
MASQUERADE    udp   --  192.168.122.0/24  !192.168.122.0/24   masq ports: 1024-65535
MASQUERADE    all   --  192.168.122.0/24  !192.168.122.0/24
```

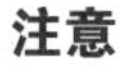

注意

不建议在虚拟交换机运行时编辑这些防火墙规则，因为有可能导致交换机通信故障。

为了简化虚拟网络中的管理，libvirt 还使用 DNSMASQ 组件为 default 网络中的虚拟机提供

DNS 和 DHCP 功能，命令如下：

```
3 # ps aux | grep dnsmasq
dnsmasq    3577  0.0  0.0    9960  2156 ?        S    10:06   0:00 /usr/sbin/dnsmasq
--conf-file=/var/lib/libvirt/dnsmasq/default.conf --leasefile-ro --dhcp-script=/usr/libexec/libvirt_leaseshelper
root       3578  0.0  0.0    9932   376 ?        S    10:06   0:00 /usr/sbin/dnsmasq
--conf-file=/var/lib/libvirt/dnsmasq/default.conf --leasefile-ro --dhcp-script=/usr/libexec/libvirt_leaseshelper
root       7668  0.0  0.0    6408  2220 pts/0    S+   10:42   0:00 grep --color=auto dnsmasq

4 # cat /var/lib/libvirt/dnsmasq/default.conf
##WARNING: THIS IS AN AUTO-GENERATED FILE. CHANGES TO IT ARE LIKELY TO BE
##OVERWRITTEN AND LOST. Changes to this configuration should be made using:
##    virsh net-edit default or other application using the libvirt API.
##
## dnsmasq conf file created by libvirt
strict-order
pid-file=/run/libvirt/network/default.pid
except-interface=lo
bind-dynamic
interface=virbr0
dhcp-range=192.168.122.2,192.168.122.254,255.255.255.0
dhcp-no-override
dhcp-authoritative
dhcp-lease-max=253
dhcp-hostsfile=/var/lib/libvirt/dnsmasq/default.hostsfile
addn-hosts=/var/lib/libvirt/dnsmasq/default.addnhosts
```

从第 3 行命令的输出中可以看出 DNSMASQ 所使用的配置文件也是由 libvirt 提供的，它的路径名为=/var/lib/libvirt/dnsmasq/default.conf。

从第 4 行命令的输出中可以看出向虚拟机分配的 IP 地址的范围是从 192.168.122.2 到 192.168.122.254。

接下来，执行如下命令：

```
5 # virsh net-list
 Name      State    Autostart   Persistent
--------------------------------------------------
 default   active   yes         yes

6 # virsh net-dumpxml default
<network>
  <name>default</name>
  <uuid>643fd5c2-1b74-4bdc-b189-690259c64a28</uuid>
  <forward mode='nat'>
    <nat>
      <port start='1024' end='65535'/>
    </nat>
  </forward>
  <bridge name='virbr0' stp='on' delay='0'/>
  <mac address='52:54:00:0f:b3:2b'/>
  <ip address='192.168.122.1' netmask='255.255.255.0'>
```

```
        <dhcp>
            <range start='192.168.122.2' end='192.168.122.254'/>
        </dhcp>
    </ip>
</network>

7 # ls /etc/libvirt/qemu/networks/
autostart    default.xml

8 # cat /etc/libvirt/qemu/networks/default.xml
<!--
WARNING: THIS IS AN AUTO-GENERATED FILE. CHANGES TO IT ARE LIKELY TO BE
OVERWRITTEN AND LOST. Changes to this xml configuration should be made using:
  virsh net-edit default or other application using the libvirt API.
-->

<network>
  <name>default</name>
  <uuid>643fd5c2-1b74-4bdc-b189-690259c64a28</uuid>
  <forward mode='nat'/>
  <bridge name='virbr0' stp='on' delay='0'/>
  <mac address='52:54:00:0f:b3:2b'/>
  <ip address='192.168.122.1' netmask='255.255.255.0'>
    <dhcp>
      <range start='192.168.122.2' end='192.168.122.254'/>
    </dhcp>
  </ip>
</network>
```

从第 5 行命令的输出可以看出，这个名为 default 的网络是随着 libvirtd 的启动而自动启动的，当前的状态是已激活。

第 6 行命令输出了 default 网络的详细定义：

(1) <name>default</name>指定了虚拟网络的名称。

(2) <uuid>643fd5c2-1b74-4bdc-b189-690259c64a28</uuid>指定了虚拟网络的全局唯一标识符。

(3) <forward mode='nat'>指定了虚拟网络连接到物理网络，mode 属性确定了转发方法。

(4) <port start='1024' end='65535'/>设置了用于<nat>的端口范围。

(5) <bridge name='virbr0' stp='on' delay='0'/>指定了 libvirt 在宿主机上创建网桥设备的信息。name 属性定义了网桥设备的名称，新网桥启用对生成树协议(STP，spanning-tree protocol)的支持，默认延迟为 0。建议使用 virbr 开头的桥名称，虚拟机连接到该桥设备，就像将真实世界的计算机连接到物理交换机一样。

(6) <mac address='52:54:00:0f:b3:2b'/> 属性定义了一个 MAC(硬件)地址，格式为 6 组 2 位十六进制数字，各组之间用冒号分隔。这个 MAC 地址在创建时即分配给桥接设备。建议让 libvirt 自动生成一个随机 MAC 地址并将其保存在配置中。

(7) <ip address='192.168.122.1' netmask='255.255.255.0'>为网桥指定 IP 地址。

(8) <dhcp>指定了在虚拟网络上启用 DHCP 服务。

(9) <range start='192.168.122.2' end='192.168.122.254'/>指定了要提供给 DHCP 客户端的地址池的边界。

提示

libvirt 网络配置的详细介绍可参考 https://libvirt.org/formatnetwork.html。

RHEL/CentOS 9 中的 libvirt 网络配置保存在/etc/libvirt/qemu/networks/目录下的 XML 文件中。强烈建议不直接编辑这些配置文件，而应通过 virsh 的 net-edit 子命令修改。

5.1.3 查看虚拟机的网络配置

下面再查看一下当启动虚拟机后网络所发生的变化，命令如下：

```
1 # virsh domiflist demo1_centos7
   Interface    Type        Source    Model     MAC
  -------------------------------------------------------------------------
   -            network     default   virtio    52:54:00:b1:a2:21

2 # virsh start demo1_centos7
  Domain 'demo1_centos7' started

3 # virsh domiflist demo1_centos7
   Interface    Type        Source    Model     MAC
  -------------------------------------------------------------------------
   vnet0        network     default   virtio    52:54:00:b1:a2:21
```

从第 1 行命令的输出中可以看出：这个名为 demo1_centos7 的虚拟机未启动时，Interface 的属性为空，它连接到网络的名称是 default。

当启动此虚拟机时，libvirt 会在宿主机上创建一个新的虚拟网络接口，并将其连接到网桥中的 virbr0 上。我们可以从系统日志(/var/log/messages)中看到类似这样的信息：

```
kernel: virbr0: port 1(vnet0) entered learning state
kernel: virbr0: port 1(vnet0) entered forwarding state
```

新的虚拟网络接口的名称是以 vnet 开头，后面是从 0 开始的序号。如果虚拟机操作系统的网络是启动获取 IP 地址的，则还会在系统日志中看到它从 DNSMASQ 中租用 IP 地址的信息：

```
dnsmasq-dhcp[3577]: DHCPDISCOVER(virbr0) 52:54:00:b1:a2:21
dnsmasq-dhcp[3577]: DHCPOFFER(virbr0) 192.168.122.106 52:54:00:b1:a2:21
dnsmasq-dhcp[3577]: DHCPREQUEST(virbr0) 192.168.122.106 52:54:00:b1:a2:21
dnsmasq-dhcp[3577]: DHCPACK(virbr0) 192.168.122.106 52:54:00:b1:a2:21
```

还可以使用 ip 和 nmcli 命令查看宿主机上网络的变化情况，命令如下：

```
4 # ip addr
……
3: virbr0: <BROADCAST,MULTICAST,UP,LOWER_UP> mtu 1500 qdisc noqueue state UP group default qlen
```

```
1000
        link/ether 52:54:00:0f:b3:2b brd ff:ff:ff:ff:ff:ff
        inet 192.168.122.1/24 brd 192.168.122.255 scope global virbr0
           valid_lft forever preferred_lft forever
  4: vnet0: <BROADCAST,MULTICAST,UP,LOWER_UP> mtu 1500 qdisc noqueue master virbr0 state
    UNKNOWN group default qlen 1000
        link/ether fe:54:00:b1:a2:21 brd ff:ff:ff:ff:ff:ff
        inet6 fe80::fc54:ff:feb1:a221/64 scope link
           valid_lft forever preferred_lft forever

  5 # nmcli connection
     NAME    UUID                                   TYPE        DEVICE
     ens160  60edd0a5-1c19-3b74-bfd5-8909037aec33   ethernet    ens160
     virbr0  d5e1fa5c-cc7b-423f-a785-01e4fd87d8b0   bridge      virbr0
     vnet0   e9e1b4a3-2bc4-4ac5-b6cf-c70aa35da882   tun         vnet0
```

从第 4 行命令的输出可以看出：新增加了一个名为 vnet0 的接口，它的 master 是 virbr0。第 5 行命令的输出显示了它的类型是 TUN，准确来讲应当是 TAP 设备。

> **注意**
>
> NetworkManager 有些版本总是将 TUN/TAP 设备显示为 TAP 设备，而 ip 命令(例如 iproute)显示的类型则是正确的。

5.1.4 libvirt 管理的虚拟网络

在 RHEL/CentOS 9 中启动 libvirtd 守护程序时，libvirtd 会读取目录/etc/libvirt/qemu/networks/autostart/中的符号链接所指向的虚拟网络配置文件，然后用 Network Manager 创建相应的网桥、TAP 等设备，并根据需要配置 IP 转发、路由表和 iptables 中的 NAT 规则，从而为虚拟机提供虚拟网络环境。

对于虚拟网络的管理，不同的管理工具也有一些细微差异，而通过 virsh 的 net-create、net-define、net-edit 和 net-update 子命令编辑网络 XML 文件的方式则支持所有的网络类型，具体在官网可以看到，官网网址为 https://libvirt.org/formatnetwork.html。

其中最常见的包括隔离模式虚拟网络、NAT 模式虚拟网络、桥接模式虚拟网络，本章也主要就这 3 种虚拟网络展开介绍。

5.2 创建和管理隔离网络

顾名思义，隔离网络是一种封闭的网络，连接在此虚拟交换机的虚拟机可以彼此通信，也可以与宿主机通信，但是它们的流量不会通过宿主机扩散到外部，当然也无法从宿主机外部访问它们。

5.2.1 通过 virsh 创建和管理隔离网络

virsh 管理虚拟网络的子命令多数是以 net 开头的，可以通过 virsh help network 获得这些子命令的清单。通过 virsh help 获得某个子命令的详细帮助，命令如下：

```
1 # virsh help network
 Networking (help keyword 'network'):
    net-autostart    autostart a network
    net-create       create a network from an XML file
    net-define       define an inactive persistent virtual network or modify an existing persistent one from an XML file
    net-destroy      destroy (stop) a network
……

2 # virsh help net-create
  NAME
    net-create - create a network from an XML file

  SYNOPSIS
    net-create <file> [--validate]

  DESCRIPTION
    Create a network.

  OPTIONS
    [--file] <string>    file containing an XML network description
    --validate           validate the XML against the schema
```

virsh 的 net-create 子命令可以根据一个 XML 文件中的设置来创建临时性的(transient)虚拟网络。net-define 子命令既可以创建新的虚拟网络，也可以修改现有的虚拟网络。net-define 子命令创建的新虚拟网络是持久的，但默认为未激活，需要使用 net-start 子命令进行激活。

下面使用 net-define 子命令来做一个示例，命令如下：

```
3 # vim isolated1.xml
<network>
  <name>isolated1</name>
  <domain name='isolated1'/>
  <ip address='192.168.100.1' netmask='255.255.255.0'>
    <dhcp>
      <range start='192.168.100.128' end='192.168.100.254'/>
    </dhcp>
  </ip>
</network>

4 # virsh net-define isolated1.xml
Network isolated1 defined from isolated1.xml

5 # virsh net-dumpxml isolated1
<network>
  <name>isolated1</name>
  <uuid>6e6538a3-7003-4ad1-ae97-71266dabed75</uuid>
```

```
    <bridge name='virbr1' stp='on' delay='0'/>
    <mac address='52:54:00:8d:06:75'/>
     <ip address='192.168.100.1' netmask='255.255.255.0'>
       <dhcp>
         <range start='192.168.100.128' end='192.168.100.254'/>
       </dhcp>
    </ip>
  </network>
```

第 3 行命令在当前目录下创建一个新文件，在其中为新的虚拟网络设置一些必要的属性。如果不指定模式，则默认为隔离模式。如果不指定 bridge 属性，则 libvirt 会根据默认值生成一个名称为 virbr 开头的网桥。

第 4 行命令定义了一个新的网桥。从第 5 行命令的输出中可以看到 libvirt 会生成一些属性值，包括 uuid、bridge、name 等。

接下来执行的命令如下：

```
6 # virsh net-list
   Name       State    Autostart    Persistent
  ------------------------------------------------------
   default    active    yes           yes

7 # virsh net-list --all
   Name          State        Autostart       Persistent
  -----------------------------------------------------------
   default       active       yes             yes
   isolated1     inactive     no              yes

8 # virsh net-start isolated1
   Network isolated1 started

9 # virsh net-autostart isolated1
   Network isolated1 marked as autostarted

10 # virsh net-list --all
   Name          State       Autostart      Persistent
  -----------------------------------------------------------
   default       active      yes            yes
   isolated1     active      yes            yes
```

由于新增加的虚拟网络并没有被激活，所以它不会出现在第 6 行命令的输出中，第 7 行命令有--all 选项，所以可以看到这个新网络。

第 8 行命令启动这个网络，第 9 行命令将其设置为自动启动。

如果需要修改虚拟网络，则既可以通过 net-edit 子命令来编辑 XML 文件，也可以通过 net-update 子命令根据另外一个 XML 的文件进行更新。

当不再需要某个虚拟网络的时候，可以先使用 net-destroy 子命令停止它，然后使用 net-undefine 子命令删除它的定义。

5.2.2 使用隔离网络

隔离模式允许虚拟机之间相互通信，但是它们无法与外部物理网络进行通信。下面通过示例进行验证。

将虚拟机 demo1_centos7 和 demo2_centos7 都更改到 isolated1 这个虚拟网络中。更改虚拟机的网络连接方法很简单，这里使用 virsh 管理工具。

使用 virsh 的 edit 子命令可以修改虚拟机的 XML 文件。例如，修改原有虚拟网卡配置，原配置如下：

```
# virsh list --all
 Id    Name                 State
----------------------------------------------------
 1      demo2_centos7    running
 -      demo1_centos7    shut off

# virsh dumpxml demo1_centos7
……
    <interface type='network'>
      <mac address='52:54:00:b1:a2:21'/>
      <source network='default'/>
      <model type='virtio'/>
      <address type='pci' domain='0x0000' bus='0x01' slot='0x00' function='0x0'/>
    </interface>
……
```

修改后的配置如下：

```
# virsh edit demo1_centos7
  ……
    <interface type='network'>
      <mac address='52:54:00:b1:a2:21'/>
      <source network='isolated1'/>
      <model type='virtio'/>
      <address type='pci' domain='0x0000' bus='0x01' slot='0x00' function='0x0'/>
    </interface>


# virsh domiflist demo1_centos7
 Interface  Type       Source     Model    MAC
-------------------------------------------------------------------------------
 -          network    isolated1  virtio   52:54:00:b1:a2:21
```

最主要的是修改<source network='isolated1'/>，修改之后重新启动虚拟机即可生效。

如果想修改正在运行的虚拟机网络卡配置，则需要使用 update-device 子命令，命令如下：

```
1 # virsh dumpxml demo2_centos7 | grep 'mac address'
      <mac address='52:54:00:1e:83:d7'/>

2 # vi new_isolated1.xml
    <interface type='network'>
      <mac address='52:54:00:1e:83:d7'/>
```

```
        <sourcc nctwork='isolated1'/>
        <model type='virtio'/>
        <address type='pci' domain='0x0000' bus='0x01' slot='0x00' function='0x0'/>
      </interface>

3 # virsh update-device --domain demo2_centos7 --file new_isolated1.xml --persistent --live
Device updated successfully

4 # virsh domiflist demo2_centos7
 Interface    Type        Source      Model    MAC
-------------------------------------------------------------------
 vnet0        network     isolated1   virtio   52:54:00:1e:83:d7
```

第 1 行命令获得 demo2_centos7 虚拟机的 mac address 信息。

第 2 行命令创建了包括新配置的 XML 文件。与原有 XML 文件的区别主要是<source network='isolated1'/>这一行。

第 3 行命令使用 XML 文件更新设备。通过--domain 指定虚拟机的名称、ID 或 uuid，--file 后可指定 XML 文件，--persistent 使其持久化，--live 用于修改正在运行的虚拟机。

从第 4 行命令的输出可以看出虚拟机的网卡切换到 isolated1 虚拟网络了。

5.3 创建和管理 NAT 网络

对于桌面虚拟化或测试环境来讲，NAT 模式是最常用的虚拟网络模式。不需要进行特别配置，此模式的虚拟机就可以访问外部网络，它还允许宿主机与虚拟机之间进行通信。NAT 模式的主要“缺点”是宿主机之外的系统无法访问虚拟机。默认的 default 网络就是 NAT 模式。

NAT 模式的虚拟网络在 iptables 的帮助下创建，其实是使用伪装选项，因此，停止 iptables 会导致虚拟机内部网络的中断。下面通过实验进行验证，命令如下：

```
1 # virsh domiflist demo1_centos7
 Interface    Type        Source     Model    MAC
-------------------------------------------------------------------
 vnet1        network     default    virtio   52:54:00:b1:a2:21

2 # virsh domifaddr demo1_centos7
 Name       MAC address          Protocol     Address
-------------------------------------------------------------------------------
 vnet1      52:54:00:b1:a2:21    ipv4         192.168.122.106/24

3 # ssh 192.168.122.106 "ping -c 2    www.baidu.com"
root@192.168.122.106's password:    输入虚拟机操作系统 root 的密码
PING www.a.shifen.com (39.156.66.18) 56(84) bytes of data.
64 bytes from 39.156.66.18 (39.156.66.18): icmp_seq=1 ttl=127 time=20.1 ms
64 bytes from 39.156.66.18 (39.156.66.18): icmp_seq=2 ttl=127 time=26.9 ms

--- www.a.shifen.com ping statistics ---
```

```
2 packets transmitted, 2 received, 0% packet loss, time 1001ms
rtt min/avg/max/mdev = 20.117/23.546/26.975/3.429 ms
```

第 1 行命令的输出结果显示虚拟机 demo1_centos7 有一个连接 default 网络的网络接口。

第 2 行命令的输出结果显示这个接口的 IP 地址是 192.168.122.106/24。

第 3 行命令的输出验证了两个结果，一个是在宿主机可以访问处于 NAT 网络的虚拟机，另外一个是从这台虚拟机可以访问宿主机外部的网络。

接下来执行的命令如下：

```
4 # systemctl stop firewalld.service

5 # iptables -L -t nat
    Chain PREROUTING (policy ACCEPT)
    target     prot opt source               destination

    Chain INPUT (policy ACCEPT)
    target     prot opt source               destination

    Chain OUTPUT (policy ACCEPT)
    target     prot opt source               destination

    Chain POSTROUTING (policy ACCEPT)
    target     prot opt source               destination

6 # ssh 192.168.122.106 "ping -c 2    www.baidu.com"
    root@192.168.122.106's password:
    PING www.a.shifen.com (39.156.66.18) 56(84) bytes of data.

    --- www.a.shifen.com ping statistics ---
    2 packets transmitted, 0 received, 100% packet loss, time 1000ms
```

第 4 行命令停止了防火墙服务。从第 5 行命令的输出中可以看出 iptables 的 NAT 表中的所有链规则均为空。

从第 6 行命令的输出结果可以看出，这台虚拟机限制无法访问宿主机外部的网络。

提示

域名解析和 DHCP 服务是由 dnsmasq 进程提供的，停止防火墙服务不会影响此进程，所以第 6 行命令的输出显示域名解析服务正常运行。

5.3.1 使用 virsh 创建 NAT 网络

在 RHEL/CentOS 9 上安装 KVM 虚拟化组件的时候，会自动创建一个名为 default 的 NAT 网络。一台宿主机上可以有多个采用 NAT 模式的虚拟网络，下面再创建一个名为 nat2 的 NAT 模式的虚拟网络。

和创建隔离网络一样，NAT 网络的创建也支持多种模式，这里仍然使用 virsh 工具。

通过 virsh 创建 NAT 模式的网络，仍然需要使用 XML 格式的配置文件，命令如下：

```
1 # vi new_nat2.xml
  <network>
    <name>nat2</name>
    <forward mod='nat'/>
    <ip address='192.168.100.1' netmask='255.255.255.0'>
      <dhcp>
        <range start='192.168.100.128' end='192.168.100.254'/>
      </dhcp>
    </ip>
  </network>

2 # virsh net-define new_nat2.xml
    Network nat2 defined from new_nat2.xml

3 # virsh net-start nat2
    Network nat2 started

4 # virsh net-autostart nat2
    Network nat2 marked as autostarted

5 # virsh net-list --all
    Name      State     Autostart    Persistent
  ----------------------------------------------------------
    default   active    yes          yes
    nat2      active    yes          yes
```

第 1 行命令在当前目录中创建了一个新文件，在其中为新的虚拟网络设置了一些必要的属性。“<forward mod='nat'/>”定义的虚拟网络是 NAT。如果不指定 bridge 属性，则 libvirt 会生成一个以 virbr 开头的网桥。

第 2 行命令定义了一个新的网络。

第 3 行命令启动这个网络。

第 4 行命令设置此网络为自动启动。

接下来指定的命令如下：

```
6 # ip address
8: virbr2: <NO-CARRIER,BROADCAST,MULTICAST,UP> mtu 1500 qdisc noqueue state DOWN group
    default qlen 1000
    link/ether 52:54:00:e4:ec:e3 brd ff:ff:ff:ff:ff:ff
    inet 192.168.100.1/24 brd 192.168.100.255 scope global virbr2
       valid_lft forever preferred_lft forever

7 # virsh net-dumpxml nat2
<network>
  <name>nat2</name>
  <uuid>3ef3efaa-5dc0-4c13-9e7d-de2f5fe0310c</uuid>
  <forward mode='nat'>
    <nat>
      <port start='1024' end='65535'/>
    </nat>
```

```
  </forward>
  <bridge name='virbr2' stp='on' delay='0'/>
  <mac address='52:54:00:e4:ec:e3'/>
  <ip address='192.168.100.1' netmask='255.255.255.0'>
    <dhcp>
      <range start='192.168.100.128' end='192.168.100.254'/>
    </dhcp>
  </ip>
</network>
```

第 6 行命令的输出显示：新增一个名为 virbr2 的网桥。

第 7 行命令会输出新网桥的详细信息，会看到有些属性使用了默认值或是自动生成的。

接下来执行的命令如下：

```
8 # iptables -L -t nat -n
    Chain PREROUTING (policy ACCEPT)
    target      prot opt source             destination

    Chain INPUT (policy ACCEPT)
    target      prot opt source             destination

    Chain OUTPUT (policy ACCEPT)
    target      prot opt source             destination

    Chain POSTROUTING (policy ACCEPT)
    target          prot opt   source           destination
    LIBVIRT_PRT     all  --    0.0.0.0/0        0.0.0.0/0

Chain LIBVIRT_PRT (1 references)
    target          prot  opt  source             destination
    RETURN          all   --   192.168.100.0/24   224.0.0.0/24
    RETURN          all   --   192.168.100.0/24   255.255.255.255
    MASQUERADE      tcp   --   192.168.100.0/24   !192.168.100.0/24    masq ports: 1024-65535
    MASQUERADE      udp   --   192.168.100.0/24   !192.168.100.0/24    masq ports: 1024-65535
    MASQUERADE      all   --   192.168.100.0/24   !192.168.100.0/24
    RETURN          all   --   192.168.122.0/24   224.0.0.0/24
    RETURN          all   --   192.168.122.0/24   255.255.255.255
    MASQUERADE      tcp   --   192.168.122.0/24   !192.168.122.0/24    masq ports: 1024-65535
    MASQUERADE      udp   --   192.168.122.0/24   !192.168.122.0/24    masq ports: 1024-65535
    MASQUERADE      all   --   192.168.122.0/24   !192.168.122.0/24

9 # route -n
Kernel IP routing table
    Destination     Gateway        Genmask          Flags   Metric   Ref    Use Iface
    0.0.0.0         192.168.1.2    0.0.0.0          UG      100      0        0 ens160
    192.168.1.0     0.0.0.0        255.255.255.0    U       100      0        0 ens160
    192.168.100.0   0.0.0.0        255.255.255.0    U       0        0        0 virbr2
    192.168.122.0   0.0.0.0        255.255.255.0    U       0        0        0 virbr0
```

libvirt 会根据 NAT 模式的虚拟网络的配置来修改 iptables 的 NAT 规则和系统的路由。

从第 8 行命令的输出可以看到 iptables 的 NAT 表的 POSTROUTING 链中有多条针对

192.168.100.0/24 的规则。正是由于这些规则才使连接到 nat2 网络中的虚拟机可以访问外部的网络。

第 9 行命令的输出显示在宿主机的路由表中，新增了一个到 192.168.100.0/24 的路由，目标地址是这个网络的数据包将通过 virbr2 交换机输送出去。这样，就可以在宿主机上访问连接到 nat2 网络的虚拟机，而且属于 default 网络的虚拟机也可以访问 nat2 网络的虚拟机。

5.3.2 使用 NAT 网络

在实验环境中，准备两台虚拟机来做实验。demo1_centos7 还属于 default 网络，将 demo2_centos7 修改为 nat2，命令如下：

```
1 # virsh edit demo2_centos7
……
    <interface type='network'>
      <mac address='52:54:00:1e:83:d7'/>
      <source network='nat2'/>
      <model type='virtio'/>
      <address type='pci' domain='0x0000' bus='0x01' slot='0x00' function='0x0'/>
    </interface>
……

2 # virsh start demo1_centos7
    Domain 'demo1_centos7' started

3 # virsh start demo2_centos7
    Domain 'demo2_centos7' started
```

第 1 行命令用于编辑虚拟配置的 XML 文件，修改网卡的属性 source，并将其修改为<source network='nat2'/>。

第 2 行、第 3 行命令分别启动了这两台属于不同虚拟机网络的虚拟机域。

接下来执行的命令如下：

```
4 # ip address
……
5: vnet0: <BROADCAST,MULTICAST,UP,LOWER_UP> mtu 1500 qdisc noqueue master virbr0 state
  UNKNOWN group default qlen 1000
    link/ether fe:54:00:b1:a2:21 brd ff:ff:ff:ff:ff:ff
    inet6 fe80::fc54:ff:feb1:a221/64 scope link
       valid_lft forever preferred_lft forever
6: vnet1: <BROADCAST,MULTICAST,UP,LOWER_UP> mtu 1500 qdisc noqueue master virbr2 state
  UNKNOWN group default qlen 1000
    link/ether fe:54:00:1e:83:d7 brd ff:ff:ff:ff:ff:ff
    inet6 fe80::fc54:ff:fe1e:83d7/64 scope link
       valid_lft forever preferred_lft forever

5 # virsh domiflist demo1_centos7
    Interface   Type      Source    Model    MAC
    -------------------------------------------------------------------------
    vnet0       network   default   virtio   52:54:00:b1:a2:21
```

```
6 # virsh domifaddr demo1_centos7
    Name         MAC address        Protocol      Address
   -------------------------------------------------------------------------
    vnet0        52:54:00:b1:a2:21    ipv4          192.168.122.106/24
7 # virsh domiflist demo2_centos7
   Interface    Type         Source    Model      MAC
   -------------------------------------------------------------------------
    vnet1       network      nat2      virtio     52:54:00:1e:83:d7

8 # virsh domifaddr demo2_centos7
    Name         MAC address        Protocol      Address
   -------------------------------------------------------------------------
    vnet1        52:54:00:1e:83:d7    ipv4          192.168.100.154/24
```

从第 4 行命令的输出可以看出宿主机上新增了两个 TAP 类型的虚拟网络接口：vnet0 和 vnet1，它们分别是网桥 virbr0 和 virbr2 的子接口。

从第 5 行至第 8 行的输出可以看出：虚拟机 demo1_centos7 连接到虚拟网络 default，它的 IP 地址是 DNSMASQ 分配的 192.168.122.106/24。虚拟机 demo2_centos7 连接到虚拟网络 nat2，它的 IP 地址是 DNSMASQ 分配的 192.168.100.154/24。

提示

宿主机上名称为 vnet 开头的 TAP 设备与虚拟机的以太网卡是“一对”设备，它们的 MAC 地址后面 5 组 2 位十六进制数字是相同的。例如 vnet0 与 demo1_centos7 的 eth0 的 MAC 地址都以“54:00:b1:a2:21”结束。

接下来，连接到虚拟机 demo2_centos7(192.168.100.154/24)来做一些测试，命令如下：

```
9 # ssh 192.168.100.154 "ping -c 2 www.baidu.com"
   root@192.168.100.154's password:
   PING www.a.shifen.com (39.156.66.18) 56(84) bytes of data.
   64 bytes from 39.156.66.18 (39.156.66.18): icmp_seq=1 ttl=127 time=26.1 ms
   64 bytes from 39.156.66.18 (39.156.66.18): icmp_seq=2 ttl=127 time=31.4 ms

   --- www.a.shifen.com ping statistics ---
   2 packets transmitted, 2 received, 0% packet loss, time 1001ms
   rtt min/avg/max/mdev = 26.173/28.794/31.416/2.626 ms
```

从第 9 行命令的输出可以看出：从宿主机可以访问到连接虚拟网络 nat2 的虚拟机，而这些虚拟机可以通过宿主机的 NAT 功能访问外部网络资源。

5.4 创建和管理桥接网络

桥接(bridge)模式的网络很常见，外部网络上的主机可以访问这种连接模式网桥中的虚拟机。

libvirt 不能直接管理桥接模式的网络，所以需要先使用 Network Manager 等工具在操作系统中创建一个网桥，然后将一个物理网卡(或捆绑在一起的多个物理网卡)分配给网桥作为子接口，最后将虚拟机连接到此网桥。

网桥是在 OSI 网络模型的第 2 层上运行的。网桥(虚拟交换机)通过 ens160 与宿主机外部的网络相连接。虚拟机 demo1_centos7 和 demo2_centos7 与宿主机在同一个子网中，外部物理网络上的主机可以检测到它们并对其进行访问。

下面来做一个桥接模式的网络实验。

5.4.1 在宿主机上创建网桥

执行的命令如下：

```
1 # ip addr
1: lo: <LOOPBACK,UP,LOWER_UP> mtu 65536 qdisc noqueue state UNKNOWN group default qlen 1000
    link/loopback 00:00:00:00:00:00 brd 00:00:00:00:00:00
    inet 127.0.0.1/8 scope host lo
       valid_lft forever preferred_lft forever
    inet6 ::1/128 scope host
       valid_lft forever preferred_lft forever
2: ens160: <BROADCAST,MULTICAST,UP,LOWER_UP> mtu 1500 qdisc mq state UP group default qlen 1000
    link/ether 00:0c:29:5d:b1:9a brd ff:ff:ff:ff:ff:ff
    altname enp3s0
    inet 192.168.1.128/24 brd 192.168.1.255 scope global noprefixroute ens160
       valid_lft forever preferred_lft forever
    inet6 fe80::20c:29ff:fe5d:b19a/64 scope link noprefixroute
       valid_lft forever preferred_lft forever
3: virbr0: <NO-CARRIER,BROADCAST,MULTICAST,UP> mtu 1500 qdisc noqueue state DOWN group
    default qlen 1000
    link/ether 52:54:00:0f:b3:2b brd ff:ff:ff:ff:ff:ff
    inet 192.168.122.1/24 brd 192.168.122.255 scope global virbr0
       valid_lft forever preferred_lft forever
```

第 1 行命令会显示创建新网桥之前的网络接口信息。IP 地址 192.168.1.128/24 与 ens160 相关联。

使用 Cockpit 来创建网桥是最简单的方法。新网桥的名称是 virbr1。将宿主机上唯一的物理网卡 ens160 做为它的子接口，如图 5-2 所示。

图 5-2　使用 Cockpit 创建新的网桥

接下来执行的命令如下：

```
2 # ip addr
1: lo: <LOOPBACK,UP,LOWER_UP> mtu 65536 qdisc noqueue state UNKNOWN group default qlen 1000
    link/loopback 00:00:00:00:00:00 brd 00:00:00:00:00:00
    inet 127.0.0.1/8 scope host lo
        valid_lft forever preferred_lft forever
    inet6 ::1/128 scope host
        valid_lft forever preferred_lft forever
2: ens160: <BROADCAST,MULTICAST,UP,LOWER_UP> mtu 1500 qdisc mq master virbr1 state UP group
    default qlen 1000
    link/ether 00:0c:29:5d:b1:9a brd ff:ff:ff:ff:ff:ff
    altname enp3s0
3: virbr0: <NO-CARRIER,BROADCAST,MULTICAST,UP> mtu 1500 qdisc noqueue state DOWN group
    default qlen 1000
    link/ether 52:54:00:0f:b3:2b brd ff:ff:ff:ff:ff:ff
    inet 192.168.122.1/24 brd 192.168.122.255 scope global virbr0
        valid_lft forever preferred_lft forever
9: virbr1: <BROADCAST,MULTICAST,UP,LOWER_UP> mtu 1500 qdisc noqueue state UP group default qlen 1000
    link/ether 00:0c:29:5d:b1:9a brd ff:ff:ff:ff:ff:ff
    inet 192.168.1.128/24 brd 192.168.1.255 scope global dynamic noprefixroute virbr1
        valid_lft 1794sec preferred_lft 1794sec
    inet6 fe80::108c:184:66aa:d3d2/64 scope link noprefixroute
        valid_lft forever preferred_lft forever

3 # nmcli connection show
   NAME    UUID                                  TYPE      DEVICE
   virbr1  0f70b535-d030-447f-af27-6abaf9c56446  bridge    virbr1
   virbr0  1d4454d0-4644-49dc-84ea-2ce25bef2689  bridge    virbr0
   ens160  60edd0a5-1c19-3b74-bfd5-8909037aec33  ethernet  ens160
```

第 2 行命令输出了创建新网桥之后网络接口的信息。IP 地址 192.168.1.128/24 与 virbr1 相关联，ens160 则称为 virbr1 的子接口。网桥 virbr1 与第一个子接口也就是 ens160 的 MAC 地址相同，都是“00:0c:29:5d:b1:9a”。

注意

如果宿主机仅有一个网卡，而且在远程通过 nmcli 或网络接口配置文件创建网桥，就要特别小心，因为错误的配置会导致无法再进行远程连接。

接下来执行的命令如下：

```
4 # cat /etc/NetworkManager/system-connections/ens160.nmconnection
    [connection]
    id=ens160
    uuid=60edd0a5-1c19-3b74-bfd5-8909037aec33
    type=ethernet
    autoconnect-priority=-999
    interface-name=ens160
```

```
    master=virbr1
    slave-type=bridge
    timestamp=1668588598
    [ethernet]
    [bridge-port]

5 # cat /etc/NetworkManager/system-connections/virbr1.nmconnection
    [connection]
    id=virbr1
    uuid=0f70b535-d030-447f-af27-6abaf9c56446
    type=bridge
    autoconnect-slaves=1
    interface-name=virbr1
    [ethernet]
    [bridge]
    stp=false

    [ipv4]
    method=auto

    [ipv6]
    addr-gen-mode=default
    method=auto
    [proxy]
```

第 4 行命令显示了网络接口 ens160 配置文件的变化。这个修改后的配置文件比之前的配置文件多出了一部分信息。其中 master=virbr1、slave-type=bridge，这两条信息就反馈出目前 ens160 是 virbr1 的接口。

第 5 行命令显示了新网桥 virbr1 的网络接口配置文件的内容，其中最重要的属性是 TYPE=bridge，而且 autoconnect-slaves=1、interface-name=virbr1，从这两条信息也能看出 ens160 和 virbr1 的关系。

5.4.2 使用网桥

以虚拟机 demo3_centos7 为例，我们通过 Cockpit 来修改器网络接口的配置。在网络接口配置中，选择接口类型为 Bridge to LAN，然后在源中选择 virbr1，单击“保存”按钮保存更改，如图 5-3 所示。

52:54:00:7c:7f:5a 虚拟网络接口设置	×
接口类型	Bridge to LAN
源	virbr1
型号	virtio (Linux, perf)
MAC 地址	52:54:00:7c:7f:5a
保存	取消

图 5-3　使用 Cockpit 更新虚拟机的网络配置

执行的命令如下：

```
1 # virsh domiflist demo3_centos7
   Interface    Type      Source    Model    MAC
  -------------------------------------------------------------------------
   vnet0        bridge    virbr1    virtio   52:54:00:7c:7f:5a

2 # virsh dumpxml demo3_centos7
   … …
       <interface type='bridge'>
          <mac address='52:54:00:7c:7f:5a'/>
          <source bridge='virbr1'/>
          <model type='virtio'/>
          <address type='pci' domain='0x0000' bus='0x01' slot='0x00' function='0x0'/>
       </interface>
   … …

3 # virsh start demo3_centos7

4 # ping -c 2 192.168.1.134
   PING 192.168.1.134 (192.168.1.134) 56(84) bytes of data.
   64 bytes from 192.168.1.134: icmp_seq=1 ttl=64 time=0.614 ms
   64 bytes from 192.168.1.134: icmp_seq=2 ttl=64 time=1.42 ms

   --- 192.168.1.134 ping statistics ---
   2 packets transmitted, 2 received, 0% packet loss, time 1040ms
   rtt min/avg/max/mdev = 0.614/1.019/1.424/0.405 ms
```

第 1 行命令的输出内容显示此虚拟机已经连接到网桥 virbr1。

第 2 行命令的输出内容显示它的网桥接口类型是<interface type='bridge'>。

第 3 行命令用于启动虚拟机。这时，虚拟机 demo3_centos7 会从宿主机所在物理网络中的 DHCP 服务器获得 IP 地址。我们可以登录到此虚拟机查看其 IP 地址，在本例中是 192.168.1.134。

第 4 行命令用于对这个 IP 地址进行测试。当然外部物理网络上的其他主机也可以访问到此虚拟机。

5.5 习题

1. 为新创建的虚拟机 new_vm 设置可用网络选项。

2. 使用 virsh 命令创建全新 NAT 模式虚拟网络，虚拟网段为 192.168.200.0/24。设置虚拟机 new_vm 使用此网络，并确保 new_vm 可以正常访问外网(例如：ping 通百度)。

3. 使用 Cockpit 工具，创建新的网桥设备，名称为 new_br，设置虚拟机 new_vm 使用此网络。

第6章

管理KVM虚拟存储

KVM能够使用Linux支持的任何存储，包括某些本地磁盘和网络附加存储(NAS)。还可以利用多路径I/O增强存储并提供冗余能力。KVM还支持共享文件系统，因此虚拟机镜像可以由多个主机共享。磁盘镜像支持精简置备，可以按需分配存储，不必预先备妥一切。

简单来说，虚拟存储，是KVM虚拟机中能够看到的“磁盘”，一般认为虚拟机是真实存在的磁盘，但其实它是宿主机物理存储的一部分，是通过模拟或者半虚拟化的块设备驱动程序分配给虚拟机的存储。

本章要点

◎ 常用存储分类
◎ qemu-img命令的使用
◎ 存储池
◎ 存储卷

6.1 常见的存储资源

常见的存储资源有很多种，如镜像文件、LVM卷、宿主机设备和分布式存储系统。

1. 镜像文件

镜像文件只能存储在宿主机文件系统中，例如本地文件系统或网络文件系统(NFS或CIFS)。

2. LVM逻辑卷

可以将LVM的逻辑卷直接分配给虚拟机，所以它提供了比镜像文件更高的性能。LVM的精简供给、快照功能可以更高效地使用存储空间。由于还可以用宿主机上的 LVM 工具进行管理，所以管理更简单。

3. 宿主机设备

可以将宿主机上的设备(例如物理 CD-ROM、磁盘、分区和 LUN)直接分配给虚拟机，从而获得媲美“原生”的性能和特性。

4. 分布式存储系统

分布式存储系统是新型的存储解决方案，它将数据分散存储在多台独立的设备上，各司其职、协同合作，统一地对外提供存储服务。这种扩展的系统结构，不但提高了系统的可靠性、可用性和存取效率，还易于扩展。KVM 目前支持 GlusterFS、Ceph 等多种分布式存储系统。

6.2 虚拟磁盘类型

KVM 中，虚拟磁盘的类型很丰富，常见的有 raw 和 qcow2 两种。要探究、学习这两种磁盘类型，就需要借助 qemu-img 磁盘管理工具，使用 qemu-img 工具可以创建镜像文件。

执行命令结果如下：

```
# qemu-img --version
    qemu-img version 7.0.0 (qemu-kvm-7.0.0-12.el9)
    Copyright (c) 2003-2022 Fabrice Bellard and the QEMU Project developers

# qemu-img --help |grep Supported
    Supported formats: blkdebug blklogwrites blkverify compresscopy-before-write copy-on-read file ftp ftps
    host_cdrom host_device http https luks nbd
    null-aio null-co nvme preallocate qcow2 quorum raw rbd snapshot-access
    throttle vdi vhdx vmdk vpc
```

镜像文件有多种格式，上面的命令列出了 qemu-img 所支持的格式，KVM 中常用的磁盘镜像格式是 raw 和 qcow2。

raw 的原意就是“未经加工”，是未经处理、也未经压缩的格式。该文件中只包含数据内容，而没有其他数据。且 raw 格式又有两种设置，一种是预分配的(pre-allocated)，另一种是稀疏的(sparse)。稀疏文件可以按需分配宿主机磁盘空间，因此它是精简配置(thin provisioning)的一种形式。预分配文件中可能会有长期“闲置”的空间，但它比稀疏文件具有更高的性能。当对磁盘 I/O 性能要求高且很少需要通过网络传输文件时，建议使用 raw 格式。但它也有缺点，就是不支持虚拟机快照功能。

qcow2 镜像文件在性能上没有 raw 格式高，但它提供了许多其他功能，包括快照、压缩、加密等，并且 qcow2 文件可以更有效地通过网络进行传输。因此在学习过程中，经常使用 qcow2 格式。

6.3 qemu-img 磁盘管理命令

qemu-img 命令是一个磁盘管理命令，这也是工作中经常用到的一条命令。

命令运行结果如下：

```
# qemu-img --help
    qemu-img version 7.0.0 (qemu-kvm-7.0.0-12.el9)
    Copyright (c) 2003-2022 Fabrice Bellard and the QEMU Project developers
    usage: qemu-img [standard options] command [command options]
    QEMU disk image utility

        '-h', '--help'          display this help and exit
        '-V', '--version'       output version information and exit
        '-T', '--trace'         [[enable=]<pattern>][,events=<file>][,file=<file>]
                                specify tracing options
    ……
```

常用选项如下：

- check：检查完整性。
- create：创建镜像。
- commit：提交更改。
- compare：比较。
- convert：转换。
- info：获得信息。
- map：映射。
- snapshot：快照管理。
- rebase：在已有的镜像基础上创建新的镜像。
- resize：调整大小。
- amend：修订镜像格式选项。

6.3.1 创建和格式化磁盘文件

创建新的磁盘镜像文件，语法格式如下：

```
# qemu-img --help |grep create
   create [--object objectdef] [-q] [-f fmt] [-b backing_file [-F backing_fmt]] [-u] [-o options] filename [size]
   contain only zeros for qemu-img to create a sparse image during
   '-n' skips the target volume creation (useful if the volume is created
   'snapshot' is the name of the snapshot to create, apply or delete
   '-c' creates a snapshot
```

以上选项有很多参数，我们可以继续再做精简。

```
# qemu-img create [-f fmt]   [-o options] filename [size]
```

释义：创建一个格式为 fmt、大小为 size、文件名为 filename 的镜像文件。根据文件格式 fmt 的不同，还可以添加一个或多个选项(options)来附加对该文件的各种功能设置，可以使用-o ?来查询某种格式文件支持哪些选项，在-o 选项中各个选项用逗号分隔。

可以通过-f 选项指定格式，默认格式是 raw，命令如下：

```
1 # qemu-img create test1.img 512M
    Formatting 'test1.img', fmt=raw size=536870912

2 # qemu-img info test1.img
    image: test1.img
    file format: raw
    virtual size: 512 MiB (536870912 bytes)
    disk size: 1 MiB
```

从第 2 行命令的输出可以看出：默认格式是 raw，而且是精简供给格式。还可以在创建磁盘的同时，加上 -f 选项，指定其格式，接下来执行如下命令：

```
3 # qemu-img create -f qcow2 test2.qcow2 512M
   Formatting 'test2.qcow2', fmt=qcow2 cluster_size=65536 extended_l2=off compression_type=zlib
     size=536870912 lazy_refcounts=off refcount_bits=16

4 # qemu-img info test2.qcow2
   image: test2.qcow2
   file format: qcow2
   virtual size: 512 MiB (536870912 bytes)
   disk size: 196 KiB
   cluster_size: 65536
   Format specific information:
           compat: 1.1
           compression type: zlib
           lazy refcounts: false
           refcount bits: 16
           corrupt: false

   extended l2: false


5 # ls -lhs test*
   1.0M -rw-r--r--. 1 root root 512M Sep 27 17:32 test1.img
   196K -rw-r--r--. 1 root root 193K Sep 28 08:18 test2.qcow2
```

第 3 行命令使用-f qcow2 选项指定了新镜像文件的格式。创建 qcow2 格式的文件时，会按默认选项进行格式化。

从第 4 行命令的输出可以看出，qcow2 格式的镜像文件属性更丰富一些。

第 5 行命令列出了两个文件的信息。在操作系统中显示 qcow2 格式的文件大小是 193KB，它并不能真实地反应其可以存储数据的空间大小(512MB)。

默认的 raw 类型是精简供给格式，可以通过 preallocation 选项来调整格式，它允许的值包括以下几个。

(1) off：默认值。不进行预分配空间，即精简(稀疏)格式。

(2) falloc：调用 posix_fallocate()函数预分配空间。

(3) full：通过将 0 写入底层存储来预分配空间。

执行以下命令对比观察：

```
6 # qemu-img create -f raw -o preallocation=full test.raw 50M
   Formatting 'test.raw', fmt=raw size=52428800 preallocation=full

7 # qemu-img info test.raw
   image: test.raw
   file format: raw
   virtual size: 50 MiB (52428800 bytes)
   disk size: 50 MiB

8 # ls -lhs *.raw
   50M -rw-r--r--. 1 root root 50M Sep 28 08:28 test.raw
```

提示

选项 falloc 是 fast allocation 的缩写。调用 posix_fallocate()函数分配文件块时，只标记但不进行初始化，创建速度比 full 模式快很多，类似于 VMware ESXi 的“厚置备延迟置零”。

6.3.2 调整磁盘文件的大小

qemu-img 命令可以增加或者减小 raw 格式的镜像文件大小，但只能增加 qcow2 镜像文件的大小而不能减小。

重新调整镜像文件大小，语法如下：

```
# qemu-img resize filename size
```

释义：改变镜像文件的大小，使其不同于创建之时的大小。

示例：

```
1 # qemu-img resize test1.img 1G
   WARNING: Image format was not specified for 'test1.img' and probing guessed raw.
   Automatically detecting the format is dangerous for raw images, write operations on block 0 will be restricted.
   Specify the 'raw' format explicitly to remove the restrictions.
   Image resized.

2 # qemu-img info test1.img
   image: test1.img
   file format: raw
   virtual size: 1 GiB (1073741824 bytes)
   disk size: 1 MiB
```

在调整大小时，建议通过-f 选项明确指定文件的格式。如果未指明，qemu-img 命令会自动判断，这对于 raw 格式的镜像文件是有风险的，所以第 1 行命令的输出中有警告信息。

另外，还可以使用相对值来调整大小，语法如下：

```
# qemu-img resize filename [ +|- ] size [ K|M|G|T ]
```

释义：+和-分别表示增加和减少镜像文件的大小，而 size 也支持 KB、MB、GB、TB 等单位的使用。

示例：

```
3 # qemu-img resize test2.qcow2 +512M
    Image resized.

4 # qemu-img info test2.qcow2
    image: test2.qcow2
    file format: qcow2
    virtual size: 1 GiB (1073741824 bytes)
    disk size: 260 KiB
    cluster_size: 65536
    Format specific information:
        compat: 1.1
        compression type: zlib
        lazy refcounts: false
        refcount bits: 16
        corrupt: false
        extended l2: false
```

注意

在减小镜像文件之前，除了需要做好备份工作之外，还必须在虚拟机中使用工具来减少文件系统和分区大小，否则可能会造成数据丢失。

6.3.3 转换镜像文件的格式

可以使用 convert 子命令转换镜像格式，语法如下：

```
# qemu-img convert -h
  convert [--object objectdef] [--image-opts] [--target-image-opts] [--target-is-zero] [--bitmaps] [-U] [-C] [-c] [-p]
        [-q] [-n] [-f fmt]
  [-t cache] [-T src_cache] [-O output_fmt] [-B backing_file [-F backing_fmt]]
  [-o options] [-l snapshot_param] [-S sparse_size] [-r rate_limit]
  [-m num_coroutines] [-W] [--salvage] filename [filename2 [...]]
  output_filename
```

简化如下：

```
# qemu-img convert [ -c] [ -p] [ -f fmt] [ -t cache ] [ -O output_fmt ]
 [ -o options] [ -S sparse_size ] filename output_filename
```

释义：将 fmt 格式的 filename 镜像文件根据 options 选项转换为格式为 output_fmt 的名为 output_filename 的镜像文件。

示例：

```
1 # qemu-img info test.raw
    image: test.raw
    file format: raw
    virtual size: 50 MiB (52428800 bytes)
    disk size: 50 MiB

2 # qemu-img convert -f raw -O qcow2 test.raw test.qcow2

3 # qemu-img info test.qcow2
    image: test.qcow2
    file format: qcow2
    virtual size: 50 MiB (52428800 bytes)
    disk size: 196 KiB
    cluster_size: 65536
    Format specific information:
        compat: 1.1
        compression type: zlib
        lazy refcounts: false
        refcount bits: 16
        corrupt: false
        extended l2: false
```

从第 1 行可以看出 test.raw 镜像文件的格式为 raw。

第 2 行的命令将 raw 格式转换为 qcow2。

第 3 行的命令进行验证查看。

提示

除了 KVM 常用格式镜像文件的转换外，此命令还支持其他常见格式之间的转换。比如可以用 VMware 平台将 vmdk 格式的文件转换为 qcow2 文件。

6.3.4 快照管理

常见的 raw 和 qcow2 格式中，qcow2 格式的文件镜像支持快照功能，因此这里的快照管理主要是对 qcow2 格式镜像的管理。语法如下：

```
# qemu-img snapshot -h
    snapshot [--object objectdef] [--image-opts] [-U] [-q]
             [-l | -a snapshot | -c snapshot | -d snapshot] filename
```

快照除了可以对虚拟机进行操作，也可以是基于 qcow2 格式的镜像文件，常用选项及参数如下：

(1) -l：列出镜像文件的所有快照。

(2) -a：应用快照(将镜像文件还原到指定的快照状态)。

(3) -c：创建快照。

(4) -d：删除快照。

示例：

```
1 # qemu-img info test.qcow2
    image: test.qcow2
    file format: qcow2
    virtual size: 50 MiB (52428800 bytes)
    disk size: 196 KiB
    cluster_size: 65536
    Format specific information:
        compat: 1.1
        compression type: zlib
        lazy refcounts: false
        refcount bits: 16
        corrupt: false
        extended l2: false

2 # qemu-img snapshot -l test.qcow2

3 # qemu-img snapshot -c snapshot1 test.qcow2

4 # qemu-img snapshot -l test.qcow2
    Snapshot list:
    ID        TAG            VM SIZE  DATE        VM CLOCK                ICOUNT
    1         snapshot1      0 B      2022-09-28  09:32:11 00:00:00.000  0

5 # qemu-img snapshot -c snapshot2 test.qcow2

6 # qemu-img snapshot -l test.qcow2
    Snapshot list:
    ID        TAG            VM SIZE    DATE          VM CLOCK                 ICOUNT
    1         snapshot1      0 B        2022-09-28    09:32:11 00:00:00.000   0
    2         snapshot2      0 B        2022-09-28    09:32:39 00:00:00.000   0

7 # qemu-img info test.qcow2
    image: test.qcow2
    file format: qcow2
    virtual size: 50 MiB (52428800 bytes)
    disk size: 388 KiB
    cluster_size: 65536
    Snapshot list:
    ID        TAG            VM SIZE    DATE          VM CLOCK                    ICOUNT
    1         snapshot1      0 B        2022-09-28    09:32:11 00:00:00.000       0
    2         snapshot2      0 B        2022-09-28    09:32:39 00:00:00.000       0
    Format specific information:
        compat: 1.1
        compression type: zlib
        lazy refcounts: false
        refcount bits: 16
        corrupt: false
        extended l2: false
```

第 1 行命令的输出显示镜像文件 test.qcow2 没有快照。

第 2 行命令加以验证，查看结果仍是没有快照。

第 3 行命令通过-c 选项指定创建名称为 snapshot1 的快照。

第 4 行命令查看验证。

第 5、第 6 行命令重复创建名称为 snapshot2 的快照，查看验证。

第 7 行通过 qemu-info 命令获得镜像文件的快照列表。

接下来执行以下命令：

```
1 # qemu-img snapshot -a 1 test.qcow2

2 # qemu-img snapshot -d snapshot2 test.qcow2

3 # qemu-img snapshot -l test.qcow2
    Snapshot list:
    ID        TAG            VM SIZE    DATE            VM CLOCK                   ICOUNT
    1         snapshot1      0          B 2022-09-28    09:32:11 00:00:00.000      0

4 # qemu-img snapshot -d snapshot1 test.qcow2

5 # qemu-img snapshot -l test.qcow2

6 # qemu-img info test.qcow2
    image: test.qcow2
    file format: qcow2
    virtual size: 50 MiB (52428800 bytes)
    disk size: 256 KiB
    cluster_size: 65536
    Format specific information:
        compat: 1.1
        compression type: zlib
        lazy refcounts: false
        refcount bits: 16
        corrupt: false
        extended l2: false
```

第 1 行命令将镜像文件恢复到快照 ID 1 的状态。

第 2、第 3 行命令删除快照 snapshot2，并查看验证。

第 4~第 6 行命令删除快照 snapshot1，并查看验证。

6.4 存储池

存储池(storage poll)是由 libvirt 管理的为虚拟机预留在宿主机上的存储容量。可以在存储池的空间中创建存储卷(storage volume)，然后将存储卷当做块设备分配给虚拟机。

由于是通过存储池和存储卷的名称将存储分配给虚拟机的，虚拟机并不需要知道存储的底层物理路径，这就提高了系统管理的灵活性。即使存储管理员和创建虚拟机的管理不是同一个人，也可以灵活地管理和使用存储。

默认 libvirt 使用基于目录的 dir 的存储池设计，/var/lib/libvirt/images 目录就是默认的存储池。

Cockpt、Virt-Manager、virsh 等工具都是通过 libvirt API 来管理虚拟存储的。其中 virsh 工具功能强大，是经常使用的工具。

virsh 有多个以 pool 开头的并与存储池有关的子命令，如表 6-1 所示。

表 6-1　virsh-pool 子命令

virsh 子命令	功能说明
pool-autostart	将存储池设置为自动启动
pool-build	构建一个存储池
pool-create	通过 XML 文件来创建并启动临时存储(无配置文件)
pool-define	通过 XML 文件定义存储池(非活动)，或者修改现有存储池
pool-delete	删除一个存储池
pool-destroy	停止一个存储池
pool-dumpxml	以 XML 格式显示存储池的信息
pool-edit	编辑存储池 XML 配置文件
pool-info	显示存储池的信息
pool-list	显示存储池列表
pool-refresh	刷新存储池
pool-start	启动一个已定义的非活动存储池
pool-undefine	取消一个不活动的存储池定义
pool-uuid	将存储池的名称转换为 UUID
pool-event	显示存储池的事件

6.4.1　查看当前存储池

查看命令如下：

```
1 # virsh pool-list --all --details
    Name     State      Autostart   Persistent   Capacity     Allocation    Available
  -------------------------------------------------------------------------------------------


    default  running    yes           yes          50.42 GiB    10.95 GiB     39.47 GiB
    iso      running    yes           yes          50.42 GiB    10.95 GiB     39.47 GiB
```

第一行命令列出了此宿主机上所有存储池的详细信息，包括活动及非活动的存储池，以及每个存储池的状态、是否自启动、是否是永久性存储池、总容量、已经分配容量及可用容量等信息。

接下来执行的命令如下：

```
2 # virsh pool-info default
```

```
        Name:              default
        UUID:              2e9419e5-158e-4dc2-bc21-cd91bce6b97d
        State:             running
        Persistent:        yes
        Autostart:         yes
        Capacity:          50.42 GiB
        Allocation:        10.96 GiB
        Available:         39.46 GiB

    3 # virsh pool-dumpxml default
        <pool type='dir'>
          <name>default</name>
          <uuid>2e9419e5-158e-4dc2-bc21-cd91bce6b97d</uuid>
          <capacity unit='bytes'>54142988288</capacity>
          <allocation unit='bytes'>11767463936</allocation>
          <available unit='bytes'>42375524352</available>
          <source>
          </source>
          <target>
            <path>/var/lib/libvirt/images</path>
            <permissions>
              <mode>0711</mode>
              <owner>0</owner>
              <group>0</group>
              <label>system_u:object_r:virt_image_t:s0</label>
            </permissions>
          </target>
        </pool>
```

第 2、第 3 行输出了存储池 default 的详细信息。其中<pool type='dir'>表示这是一个基于目录的存储池。

接下来执行的命令如下：

```
    4 # tree /etc/libvirt/storage/
       /etc/libvirt/storage/
       ├── autostart
       │   ├── default.xml -> /etc/libvirt/storage/default.xml
       │   └── iso.xml -> /etc/libvirt/storage/iso.xml
       ├── default.xml
       └── iso.xml

       1 directory, 4 files
```

每个存储池的 XML 文件都保存在/etc/libvirt/storage/目录中，不建议直接修改此目录中的文件。第 4 行命令显示了此目录的文件及子目录。如果在 autostart 子目录有同名的符号连接文件，则说明它是自动启动的存储池。

6.4.2 存储池分类

从不同的视角可以将存储池划分成不同的类。

1. 根据存储池的生命周期划分

(1) 持久存储池：在宿主机重新引导后仍然有效。

(2) 临时存储池：仅在宿主机当前运行时有效，重新引导之后会消失。

2. 根据底层存储技术来划分

(1) 基于目录的存储池。

(2) 基于磁盘的存储池。

(3) 基于分区的存储池。

(4) 基于 GlusterFS 的存储池。

(5) 基于 iSCSI 的存储池。

(6) 基于 LVM 的存储池。

(7) 基于 NFS 的存储池。

(8) 基于多路径的存储池。

3. 根据数据存储的位置划分

(1) 本地存储池：使用的是直接连接到宿主机上的存储。

(2) 网络(共享)存储池：通过标准协议使用网络上共享的存储设备。

6.4.3 创建基于目录的存储池

基于目录的存储池是最常见的存储池类型。它的容器就是文件系统的目录，其中的存储卷就是目录中的镜像文件。

创建一个 XML 文件，可以直接使用 vim 命令新建，可以使用 virsh 命令生产，也可以复制其他存储池的 XML 文件，修改指定参数。

复制默认存储池 XML，修改指定参数，用于指定新存储池参数，命令如下：

```
1 # cd /etc/libvirt/storage/
2 # cp -rp default.xml new.xml
3 # vim new.xml
    <pool type='dir'>
      <name>vm2</name>
      <target>
              <path>/var/lib/libvirt/images/vm2</path>
      </target>
    </pool>
```

具体 new.xml 文件中参数的含义，如表 6-2 所示。

表 6-2 new.xml 文件中参数的含义

XML 格式	描述
<pool type='dir'>	存储池类型
<name>	存储池名称
<target>	存储池目标路径

接下来，使用 virsh，根据上述 new.xml 文件生成存储池，命令如下：

```
4 # virsh pool-define new.xml
    Pool vm1 defined from new.xml

5 # virsh pool-list --all
    Name        State       Autostart
  -------------------------------------------
  default       active      yes
    iso         active      yes
    vm1         inactive    no
```

上述命令中，第 4 行命令仅仅创建了新存储池的定义。从第 5 行命令的输出可以看出：新存储池 vm1 并没有启动，而是处于“非活动状态”。

也可以使用 pool-define-as 子命令创建存储池。这种方式比较适合在命令行或脚本中使用，命令如下：

```
# virsh pool-define-as --name vm2 --type dir --target "/vm2"
    Pool vm2 defined
```

接下来执行如下命令：

```
6 # ls -t /etc/libvirt/storage/
    vm2.xml   vm1.xml   new.xml   autostart   iso.xml   default.xml

7 # cat /etc/libvirt/storage/vm2.xml
   <!--
   WARNING: THIS IS AN AUTO-GENERATED FILE. CHANGES TO IT ARE LIKELY TO BE
   OVERWRITTEN AND LOST. Changes to this xml configuration should be made using:
    virsh pool-edit vm2 or other application using the libvirt API.
   -->

   <pool type='dir'>
         <name>vm2</name>
        <uuid>e406c918-e5f8-4e87-aa0c-2eb458a34f70</uuid>
         <capacity unit='bytes'>0</capacity>
        <allocation unit='bytes'>0</allocation>
        <available unit='bytes'>0</available>
         <source>
        </source>
         <target>
           <path>/vm2</path>
        </target>
   </pool>
```

第 6 行命令的输出显示，在/etc/libvirt/storage/目录中新增了一个与存储池同名的 XML 文件，第 7 行命令显示了该 XML 文件的内容。

接下来执行以下命令：

```
8 # ls -ld /vm2
    ls: cannot access '/vm2': No such file or directory
```

```
9 # virsh pool-build vm2
    Pool vm2 built

10 # ls -ld /vm2
    drwx--x--x. 2 root root 6 Oct 17 09:07 /vm2
```

当通过 pool-build 子命令构建存储池时，如果目标目录结构(/vm2)不存在，libvirt 才会创建目标目录。

接下来执行以下命令：

```
11 # virsh pool-list --all
     Name        State        Autostart
    ----------------------------------------
    default      active       yes
    iso          active       yes
    vm1          active       yes
    vm2          inactive     no

12 # virsh pool-start vm2
    Pool vm2 started

13 # virsh pool-autostart vm2
    Pool vm2 marked as autostarted

14 # virsh pool-list --all
      Name       State       Autostart
    ----------------------------------------
      default    active      yes
      iso        active      yes
      vm1        active      yes
      vm2        active      yes
```

第 12 行命令用于启动存储池 vm2。只有当存储池启动之后，虚拟机才能访问其中的存储卷。

第 13 行命令将存储池 vm2 设置为自动启动。

第 14 行命令显示结果，用于验证第 12、第 13 行命令的内容。

当没有虚拟机使用存储池中的存储卷时，就可以将其删除，命令如下：

```
15 # virsh pool-destroy vm2
    Pool vm2 destroyed

16 # virsh pool-undefine vm2
    Pool vm2 has been undefined

17 # virsh pool-list --all
      Name       State       Autostart
    ----------------------------------------
    default      active      yes
     iso         active      yes
```

```
    vm1        active     yes
```

第 15 行命令用于停止存储池 vm2，它将处于 inactive 状态。

第 16 行命令取消了这个不活动的存储池的定义，同时也会删除/etc/libvirt/storage/目录中对应的 XML 文件。

6.4.4 创建基于 LVM 逻辑卷的存储池

逻辑卷管理是 Linux 上最灵活、使用最广泛的存储管理技术。基于 LVM 卷组的存储池的容器是卷组，其中的存储卷就是 LVM 逻辑卷。

libvirt 既可以使用现有的 LVM 卷组构建存储池，也可以通过权限的 LVM 卷组实现存储池。

下面将通过 virsh 创建使用新卷组的存储池，它是由两块物理磁盘组成的 LVM 卷组，命令如下：

```
1 # parted -l
    Error: /dev/sdb: unrecognised disk label
    Model: VMware, VMware Virtual S (scsi)
    Disk /dev/sdb: 107GB
    Sector size (logical/physical): 512B/512B
    Partition Table: unknown
    Disk Flags:

    Error: /dev/sdc: unrecognised disk label
    Model: VMware, VMware Virtual S (scsi)
    Disk/dev/sdc: 107GB
    Sector size (logical/physical): 512B/512B
    Partition Table: unknown
    Disk Flags:
```

/dev/sdb 和/dev/sdc 是两个未使用的 107GB 的磁盘。

接下来执行以下命令：

```
2 # vim new.xml
    <pool type='logical'>
      <name>vm3</name>
        <source>
          <device path = '/dev/sdb'/>
          <device path = '/dev/sdc'/>
    </source>
    <target>
          <path>/dev</path>
        </target>
    </pool>
```

第 2 行命令创建了一个新 XML 文件，它保存了新存储池的参数，参数说明如表 6-3。

表 6-3　基于 LVM 卷组的存储池的参数

XML 格式	描述
<pool type='logical'>	存储池的类型
<name>name</name>	存储池的名称
<source><device path='device_path'/>	源存储设备的路径
<name>VGname</name>	LVM 卷组的名称
<format type='lvm2'/></source>	LVM 卷组的格式
<target><path=target_path/></target>	存储池的目标路径

如果需要创建全新的 LVM 卷组，则可以通过多个<device path='device_path'/>指定多个物理卷，它们既可以是磁盘，也可以是分区。如果使用现有的 LVM 卷组，则需要通过<name>VGname</name>指定卷组名称，通过<format type='lvm2'/>指定 LVM 版本。

接下来执行如下命令：

```
3 # virsh pool-define new.xml
    Pool vm3 defined from new.xml

4 # virsh pool-list --all --details
     Name   State      Autostart   Persistent   Capacity     Allocation   Available
    -----------------------------------------------------------------------------------------
    default  running    yes         yes          50.42 GiB    11.10 GiB    39.33 GiB
    iso      running    yes         yes          50.42 GiB    11.10 GiB    39.33 GiB
    vm1      running    yes         yes          50.42 GiB    11.10 GiB    39.33 GiB
    vm3      inactive   no          yes          -            -            -
```

第 3 行命令创建了存储池的定义。

从第 4 行命令的输出可以看出：新的存储池 vm3 处于非活动状态。由于还没有初始化，所以未显示容量信息。

接下来执行如下命令：

```
5 # virsh pool-build vm3
    Pool vm3 built

6 # pvs
   PV               VG       Fmt    Attr   PSize      PFree
   /dev/nvme0n1p2   cs_192   lvm2   a--    <79.00g    0
   /dev/sdb         vm3      lvm2   a--    <100.00g   <100.00g
   /dev/sdc         vm3      lvm2   a--    <100.00g   <100.00g

7 # vgs
   VG       #PV #LV #SN Attr    VSize      VFree
   cs_192    1   3   0  wz--n-  <79.00g    0
   vm3       2   0   0  wz--n-  199.99g    199.99g
```

第 5 行命令将构建存储池。由于这是基于 LVM 卷组的存储池，所以 libvirt 会创建 LVM 的物理卷并组成卷组。

从第 6 行命令的输出可以看出：新增了两个物理卷(/dev/sdb 和/dev/sdc)，它们都属于卷组 vm3。

第 7 行命令输出了宿主机上 LVM 卷组的信息。

接下来执行如下命令：

```
8 # virsh pool-start vm3
  Pool vm3 started

9 # virsh pool-autostart vm3
  Pool vm3 marked as autostarted

10 # virsh pool-list --all --details
  Name    State     Autostart   Persistent   Capacity      Allocation   Available
  ---------------------------------------------------------------------------------------
  default running   yes         yes          50.42 GiB     11.10 GiB    39.33 GiB
  iso     running   yes         yes          50.42 GiB     11.10 GiB    39.33 GiB
  vm1     running   yes         yes          50.42 GiB     11.10 GiB    39.33 GiB
  vm3     running   yes         yes          199.99 GiB    0.00 B       199.99 GiB
```

第 8 行命令用于启动存储池，第 9 行命令又将其设置为自动启动。

第 10 行命令显示当前存储池的列表。新存储池 vm3 的容量就是 LVM 卷组的容量。

删除基于 LVM 卷组的存储池时，还是先使用 pool-destroy 子命令停止该存储池，然后使用 pool-undefine 子命令取消存储池的定义。

需要注意的是：删除基于 LVM 卷组的存储池，并不会自动删除 LVM 的卷组、逻辑卷及数据。

6.4.5 创建基于网络文件系统的存储池

在生产环境中，通常应将虚拟机的存储保存在集中的外部存储设备上，而不是放置在宿主机的本地存储中。基于网络文件系统的存储池就是一种常见的集中存储解决方案。

常用的网络文件系统有 NFS 和 CIFS，一旦将这些共享的存储卷挂载到宿主机的目录上，其管理操作与本地文件系统就类似。下面以 NFS 来做实验。

在实验中，部署一台 CentOS 8 的主机(名称为 stor1，IP 地址是 192.168.1.201)。在其中安装 NFS 服务器端组件，它为虚拟化宿主机提供了 NFS 共享，操作命令如下：

```
1 # cat /etc/redhat-release
  CentOS Linux release 8.4.2105

2 # yum install nfs-utils

3 # systemctl enable nfs-server

4 # mkdir /vmdata

5 # chmod +x /vmdata

6 # echo "/vmdata *(rw,no_root_squash,sync)" >> /etc/exports
```

```
7 # systemctl restart nfs-server

8 # showmount -e
    Export list for stor1:
    /vmdata *
```

NFS 服务器端组件包含在 nfs-utils 软件包中，通过第 2 行命令进行安装。将/vmdata 作为共享的目录，并设置适当的权限。

第 8 行命令进行自我检测，检测 NFS 服务器是否配置正确。

需要注意，此时 NFS 服务器端不考虑防火墙和 SElinux。如果防火墙是开启状态，需要使用 firewall-cmd 命令开启 NFS 入站流量。更严谨的配置应该是 NFS 权限及防火墙规则，仅允许来自虚拟化宿主机的访问。

在创建存储池之前，要保证虚拟化宿主机可以正确地访问 NFS 共享目录，下面做一些简单的测试，命令如下：

```
9 # showmount -e 192.168.1.201
    Export list for 192.168.1.201:
    /vmdata *

10 # mkdir /testfs

11 # mount -t nfs 192.168.1.201:/vmdata /testfs/

12 # echo "test" > /testfs/1.txt

13 # cat /testfs/1.txt
test

14 # cat /vmdata/1.txt
test

15 # rm /testfs/1.txt

16 # umount /testfs

17 # rmdir /testfs/
```

注意以上命令中，第 14 行命令是在 NFS 服务器端查看对应共享目录。

测试通过之后就可以创建 XML 文件，它将用于设置新存储池的参数。参数的含义如表 6-4 所示，执行命令如下：

```
18 # vim new.xml
    <pool type='netfs'>
    <name>vm4</name>
    <source>
    <format type = "auto"/>
    <host name ="192.168.1.201"/>
    <dir path = "/vmdata"/>
```

```
</source>
<target>
<path>/vm4</path>
</target>
</pool>
```

表 6-4　基于网络文件系统的存储池参数

XML 格式	描述
<pool type="netfs">	存储池的类型
<name>name</name>	存储池的名称
<source><hostname=hostname/>	网络文件服务器的主机名或 IP 地址
<format type="auto"/> 也可以是具体的值： <format type="nfs"/> <format type="glusterfs"/> <format type="cifs"/>	存储池的格式
<dir path=source_path/><source>	网络文件服务器上的共享目录
<target> <path>target_path </path> </target>	存储池的目标路径

在 XML 定义中，通过<target><path>/vm4</path></target>指定宿主机目标路径，如果路径目录下不存在，则进行 pool-build 时会自动创建，命令如下：

```
19 # virsh pool-define new.xml
   Pool vm4 defined from new.xml

20 # virsh pool-build vm4
   Pool vm4 built

21 # virsh pool-list --all --details
  Name     State     Autostart   Persistent   Capacity    Allocation   Available
  --------------------------------------------------------------------------------
  default  running   yes         yes          50.42 GiB   11.10 GiB    39.33 GiB
  iso      running   yes         yes          50.42 GiB   11.10 GiB    39.33 GiB
  vm1      running   yes         yes          50.42 GiB   11.10 GiB    39.33 GiB
  vm4      inactive  no          yes          -           -            -
```

第 19 行命令创建存储池 vm4 的定义。

第 20 行命令构建存储池。libvirt 会坚持目标路径/vm4，如果不存在，则会创建新目录。

接下来执行以下命令：

```
21 # virsh pool-start vm4
   Pool vm4 started
```

```
22 # virsh pool-autostart vm4
   Pool vm4 marked as autostarted

23 # virsh pool-list --all --details
   Name     State     Autostart   Persistent   Capacity    Allocation   Available
   ----------------------------------------------------------------------------------
   default  running   yes         yes          50.42 GiB   11.10 GiB    39.33 GiB
   iso      running   yes         yes          50.42 GiB   11.10 GiB    39.33 GiB
   vm1      running   yes         yes          50.42 GiB   11.10 GiB    39.33 GiB
   vm4      running   yes         yes          36.95 GiB   1.77 GiB     35.18 GiB

24 # mount | grep 192.168.1.201
   192.168.1.201:/vmdata on /vm4 type nfs4
   (rw,nosuid,nodev,noexec,relatime,vers=4.2,rsize=262144,wsize=262144,namlen=255,hard,proto=tcp,timeo=
   600,retrans=2,sec=sys,clientaddr=192.168.1.128,local_lock=none,addr=192.168.1.201)
```

第 21 行命令用于启动存储池，libvirt 会将网络文件服务器的共享目录/vmdata 挂载到宿主机的目标目录/vm4 上。这可以从第 24 行命令得到验证。

第 22 行命令将存储池 vm4 加入自动启动。第 23 行命令显示了当前存储池的列表。

与基于目录的存储池类似，基于网络文件系统的存储池也是挂载目录，其中的存储卷就是此目录中的镜像文件。删除基于网络文件系统的存储池的方法也类似。

6.5 存储卷

可以将存储池划分为存储卷(storage volume)。存储卷可以是镜像文件、物理分区、LVM 逻辑卷或其他可以被 libvirt 管理的存储。无论底层硬件架构如何，存储卷都将作为块存储设备呈现给虚拟机。

virsh 有多个以 vol 开头的并与存储卷有关的子命令，如表 6-5 所示。

表 6-5 virsh 中存储卷管理子命令

virsh 子命令	功能说明
vol-clone	根据现有卷克隆出新卷
vol-create-as	根据参数创建新卷
vol-create	根据 XML 文件创建新卷
vol-create-from	使用另外一个卷作为输入创建新卷
vol-delete	删除卷
vol-download	将卷中内容下载到文件
vol-dumpxml	以 XML 格式显示卷的详细信息
vol-info	显示卷的基本信息
vol-key	根据卷名或路径返回卷的键值

(续表)

virsh 子命令	功能说明
vol-list	显示存储中卷的列表
vol-name	根据卷键值或路径返回卷的名称
vol-path	根据卷名或键值返回卷的路径
vol-pool	根据卷键值或路径返回卷
vol-resize	调整卷大小
vol-upload	将文件内容上传到卷
vol-wipe	将文件内容擦除掉

6.5.1 查看存储卷信息

查看存储卷信息有一个前提，此命令的查看范围是某一存储池中的卷，因此需要在命令后加上存储池，语法如下：

```
# virsh vol-list <pool_name> [--details]
```

在查看卷信息时，使用--details 选项会额外显示卷类型、总容量、已分配容量等信息。

举例查看默认(default)存储池中卷的列表，执行的命令如下：

```
1 # virsh pool-list
   Name        State      Autostart
  -------------------------------------------
   default     active     yes
   iso         active     yes
   vm1         active     yes


2 # virsh vol-list default
   Name                         Path
  ------------------------------------------------------------------------------
   demo1_centos7.qcow2          /var/lib/libvirt/images/demo1_centos7.qcow2
   vm2                          /var/lib/libvirt/images/vm2
```

第 1 行命令显示出存储池列表。

第 2 行命令输出显示了存储池 default 中的存储卷列表。

想要查看某个卷的信息，可以使用子命令 vol-info，命令如下：

```
3 # virsh vol-info --pool default --vol demo1_centos7.qcow2
  Name:           demo1_centos7.qcow2
  Type:           file
  Capacity:       10.00 GiB
  Allocation:     1.33 GiB
```

第 3 行命令输出显示存储池 default 中 demo1_centos7.qocw2 卷的基本信息。

libvirt 还支持用其他方式查看卷的信息。比如用来返回卷的路径，命令如下：

```
4 # virsh vol-path --pool default --vol demo1_centos7.qcow2
    /var/lib/libvirt/images/demo1_centos7.qcow2
```

接下来运行以下命令：

```
5 # virsh vol-dumpxml --pool default --vol demo1_centos7.qcow2
  <volume type='file'>
    <name>demo1_centos7.qcow2</name>
    <key>/var/lib/libvirt/images/demo1_centos7.qcow2</key>
    <capacity unit='bytes'>10737418240</capacity>
    <allocation unit='bytes'>1427124224</allocation>
    <physical unit='bytes'>10739318784</physical>
    <target>
      <path>/var/lib/libvirt/images/demo1_centos7.qcow2</path>
      <format type='qcow2'/>
      <permissions>
        <mode>0600</mode>
        <owner>107</owner>
        <group>107</group>
        <label>system_u:object_r:svirt_image_t:s0:c10,c329</label>
      </permissions>
      <timestamps>
        <atime>1666249247.088882142</atime>
        <mtime>1664198700.159151927</mtime>
        <ctime>1664198700.159151927</ctime>
        <btime>0</btime>
      </timestamps>
      <compat>1.1</compat>
      <clusterSize unit='B'>65536</clusterSize>
      <features>
        <lazy_refcounts/>
      </features>
    </target>
  </volume>
```

第 5 行命令以 XML 文件格式输出了存储卷 demo1_centos7.qcow2 的详细信息。

6.5.2 创建存储卷

由于存储池技术已经将底层的存储硬件进行了抽象，所以在不同类型的存储池中创建卷的操作基本相同。根据创建新存储卷参数的由来划分，virsh 有 4 种创建存储卷的方法。

(1) 根据参数创建新卷：vol-create-as。

(2) 根据 XML 文件创建新卷：vol-create。

(3) 根据现有卷创建新卷：vol-create-from。

(4) 根据现有卷克隆出新卷：vol-clone。

1. 根据参数创建新卷

子命令 vol-create-as 可以根据命令行参数创建新卷。其语法格式如下：

```
# vol-create-as <pool> <name> <capacity> [--allocation <string>] [--format <string>] [--backing-vol <string>]
```

```
[--backing-vol-format <string>] [--prealloc-metadata] [--print-xml]
```

(1) [--pool] <string>：必须参数，指定存储池的名称。

(2) [--name] <string>：必须参数，指定新存储卷的名称。

(3) [--capacity] <string>：必须参数，指定存储卷的大小，以整数表示。可以使用的后缀有 b、k、M、G、T，分别表示字节、千字节、兆字节、千兆字节和太字节。默认为字节。

(4) --allocation <string>：指定初始分配大小，以整数表示。默认为字节。

(5) --format <string>：指定文件格式类型。仅使用在基于文件的存储池，可接收的类型包括 raw、bochs、qcow、qcow2、qed、host_device 和 vmdk。默认格式是 raw。

(6) --backing-vol <string>：指定基础卷。创建快照时，会使用此项。

(7) --backing-vol-format <string>：指定基础卷的格式。创建快照时，会使用此项。

(8) --prealloc-metadata：预分配元数据。

根据参数创建新卷的命令如下：

```
1 # virsh vol-create-as --pool vm2 --name test1.qcow2 --capacity 500M --format qcow2
    Vol test1.qcow2 created

2 # virsh vol-list vm2 --details
    Name              Path                   Type   Capacity     Allocation
   -------------------------------------------------------------------------------
    test1.qcow2       /vm2/test1.qcow2       file   500.00 MiB   196.00 KiB

3 # qemu-img info /vm2/test1.qcow2
   image: /vm2/test1.qcow2
   file format: qcow2
   virtual size: 500 MiB (524288000 bytes)
   disk size: 196 KiB
   cluster_size: 65536
   Format specific information:
       compat: 0.10
       compression type: zlib
       refcount bits: 16
```

第 1 行命令在存储池 vm2 中创建了一个新卷 test1.qocw2，其容量为 500MB，格式为 qcow2。第 2、第 3 行命令显示了新存储卷的信息。

2. 根据 XML 文件创建新卷

子命令 vol-create 根据包含存储卷参数的 XML 文件创建一个新的存储卷。其语法格式如下：

```
# virsh vol-create <pool> <file> [--prealloc-metadata]
```

在一个基于目录的存储池中创建新的存储卷，命令如下：

```
1 # vim test2.xml
   <volume>
     <name>test2.qcow2</name>
     <capacity>1073741824</capacity>
     <allocation>0</allocation>
     <target>
```

```
        <format type='qcow2'/>
    </target>
  </volume>

2 # virsh vol-create vm2 test2.xml
    Vol test2.qcow2 created from test2.xml


3 # virsh vol-list vm2 --details
   Name          Path                 Type   Capacity     Allocation
  -----------------------------------------------------------------------------
   test1.qcow2   /vm2/test1.qcow2     file   500.00 MiB   196.00 KiB
   test2.qcow2   /vm2/test2.qcow2     file   1.00 GiB     196.00 KiB

4 # qemu-img info /vm2/test2.qcow2
  image: /vm2/test2.qcow2
  file format: qcow2
  virtual size: 1 GiB (1073741824 bytes)
  disk size: 196 KiB
  cluster_size: 65536
  Format specific information:
      compat: 0.10
      compression type: zlib
      refcount bits: 16
```

第 1 行命令创建了一个新的 XML 文件，它保存了新存储卷的参数。新卷的名称为 test2.qcow2，容量是 1GB，格式为 qcow2。

第 2 行命令用于创建新的存储卷。

第 3、第 4 行显示了新存储卷的信息。镜像文件的路径为/vm2/test2.qocw2。

3. 根据现有卷创建新卷

子命令 vol-create-from 与 vol-create 类似，用于创建新的存储卷。不过它在创建新卷时会以一个新卷作为输入，这样新卷具有与现有卷相同的数据。其语法格式如下：

```
# virsh vol-create-from <pool> <file> <vol> [--inputpool <string>] [--prealloc-metadata] [--reflink]
```

(1) --pool <string>：必须参数，指定新存储卷所在的存储池名称或者 UUID。该存储池不必与现有存储卷的存储池相同。

(2) --file <string>：必须参数，指定包含存储卷参数的 XML 文件。

(3) --vol <string>：必须参数，指定现有存储卷的名称或路径。

(4) --inputpool <string>：可选参数，指定现有存储卷所在的存储池名称。

(5) --prealloc-metadata：可选参数，设置预分配元数据。

根据现有卷创建新卷的命令如下：

```
1 # cat test2.xml
  <volume>
    <name>test2.qcow2</name>
    <capacity>1073741824</capacity>
```

```
    <allocation>0</allocation>
    <target>
      <format type='qcow2'/>
    </target>
  </volume>

2 # virsh vol-create-from --pool vm1 --file test2.xml /vm2/test2.qcow2
    Vol test2.qcow2 created from input vol test2.qcow2

3 # virsh vol-info --pool vm1 --vol test2.qcow2
  Name:           test2.qcow2
  Type:           file
  Capacity:       1.00 GiB
  Allocation:     196.00 KiB
```

第 1 行命令创建了新卷的定义。

第 2 行命令会在存储池 vm1 中创建一个名为 test2.qcow2 的新卷，新卷中的数据来自/vm2/test2.qcow2。

第 3 行命令显示了新存储卷的信息。

4. 根据现有卷克隆新卷

子命令 vol-clone 是 vol-create-from 的一个简化的、易用的版本。它会在相同的池中创建一个现有卷的副本。其语法格式如下：

```
# virsh vol-clone <vol> <newname> [--pool <string>] [--prealloc-metadata] [--reflink]
```

由于是克隆，所以无须为新卷指定大小、格式等属性。

下面把存储池 vm2 中的存储卷 test1.qcow2 克隆一份，新卷名称为 test3.qcow2，命令如下：

```
1 # virsh vol-list vm2
   Name           Path
  ---------------------------------
   test1.qcow2    /vm2/test1.qcow2
   test2.qcow2    /vm2/test2.qcow2

2 # virsh vol-clone --pool vm2 test1.qcow2 test3.qcow2
    Vol test3.qcow2 cloned from test1.qcow2

3 # virsh vol-list vm2 --details
   Name           Path               Type    Capacity     Allocation
  ------------------------------------------------------------------------
   test1.qcow2    /vm2/test1.qcow2   file    500.00 MiB   196.00 KiB
   test2.qcow2    /vm2/test2.qcow2   file    1.00 GiB     196.00 KiB
   test3.qcow2    /vm2/test3.qcow2   file    500.00 MiB   196.00 KiB
```

第 1 行命令查看当前存储池 2 中的存储卷信息。

第 2 行命令从存储池 vm2 中，复制 test1.qcow2 存储卷，并将新的存储卷命名为 test3.qcow2。

第 3 行命令中，再次查看存储池 vm2 中的存储卷信息，因为存储卷 test3.qcow2 是从 test1.qcow2 复制而来的，所以这两个存储卷的信息一致。

6.5.3 存储卷管理

1. 向虚拟机分配存储卷

virsh 中有 3 个子命令可以用来向虚拟机分配存储卷。

(1) edit：通过编辑虚拟机的 XML 文件来管理存储池。这种方法工作量大、易出错。

(2) attach-device：通过 XML 文件向虚拟机添加包括存储设备在内的新设备。

(3) attach-disk：通过参数向虚拟机添加新的存储设备，可以认为它是 attach-device 的一种再封装简化版本。

attach-device 子命令示例如下：

```
# virsh attach-device --help
    NAME
      attach-device - attach device from an XML file

    SYNOPSIS
      attach-device <domain> <file> [--persistent] [--config] [--live] [--current]

    DESCRIPTION
      Attach device from an XML <file>.

    OPTIONS
      [--domain] <string>   domain name, id or uuid
      [--file] <string>     XML file
      --persistent          make live change persistent
      --config              affect next boot
      --live                affect running domain
      --current             affect current domain
```

对主要参数说明如下：

(1) --domain：指定虚拟机的名称(或 ID、UUID)。

(2) --file：指定 XML 文件。

(3) --config：指定下次启动后生效。本次修改的是配置文件。

(4) --live：修改正在运行的虚拟机。

(5) --current：如果虚拟机正在运行，就修改当前的虚拟机，相当于--live。如果虚拟机没有运行，就修改配置文件，相当于--config。

接下来执行的命令如下：

```
1 # vi new.xml
   <disk type='volume' device = 'disk'>
            <driver name = 'qemu' type = 'qcow2'/>
            <source pool = 'vm2' volume = 'test1.qcow2'/>
            <target dev = 'vdb' bus = 'virtio'/>
   </disk>

2 # virsh attach-device demo1_centos7 new.xml --config
   Device attached successfully
```

```
3 # virsh domblklist demo1_centos7
  Target    Source
-----------------------------------------------------
  vda       /var/lib/libvirt/images/demo1_centos7.qcow2
  vdb       test1.qcow2
  sda       -
```

第 1 行命令用于创建 XML 文件。通过<source pool = 'vm2' volume = 'test1.qcow2'/>指定存储池和存储卷信息。

第 2 行命令根据 XML 向虚拟机 demo1_centos7 添加新磁盘。

第 3 行命令的输出显示，test1.qcow2 镜像文件已被成功添加到虚拟机 demo1_centos7。

2. 删除存储卷及擦除存储卷

vol-delete 子命令会从存储池中删除存储卷信息。根据存储池类型的不同，删除操作的结果会有所差异，例如：基于目录的存储池是删除镜像文件，而基于 LVM 的存储池是删除逻辑卷。

删除基于目录的存储池中的存储卷，命令如下：

```
1 # virsh vol-list vm2
  Name           Path
--------------------------------
  test1.qcow2    /vm2/test1.qcow2
  test2.qcow2    /vm2/test2.qcow2
  test3.qcow2    /vm2/test3.qcow2

2 # virsh vol-delete --pool vm2 --vol test3.qcow2
  Vol test3.qcow2 deleted

3 # virsh vol-list vm2
  Name           Path
--------------------------------
  test1.qcow2    /vm2/test1.qcow2
  test2.qcow2    /vm2/test2.qcow2
```

第 1 行命令显示了存储池 vm2 中的存储卷列表。

第 2 行命令用于删除指定卷。将存储池 vm2 中，名称为 test3.qcow2 的存储卷删除。

第 3 行命令显示、验证删除结果。

使用 vol-delete 子命令删除存储卷是一种“逻辑”删除，有可能通过一些恢复工具来恢复部分或者全部数据，所以建议在生产环境中先擦除数据再进行删除。

使用 vol-wipe 子命令可以擦除存储卷中的数据，这样可以避免恢复先前的数据。在使用时，可以通过--algorithm 选项指定擦除算法。

接下来要删除存储池 vm2 中的存储卷 test2.qcow2，运行命令如下：

```
4 # virsh vol-wipe --pool vm2 --vol test2.qcow2
  Vol test2.qcow2 wiped

5 # virsh vol-delete --pool vm2 --vol test2.qcow2
  Vol test2.qcow2 deleted
```

```
6 # virsh vol-list vm2
   Name              Path
  ---------------------------------
   test1.qcow2       /vm2/test1.qcow2
```

第 4 行命令先擦除存储卷的数据，然后执行第 5 行命令进行删除。

第 6 行命令显示、验证删除结果。

6.6 习题

1. 总结并熟悉存储池、存储卷的相关命令。
2. 创建基于目录的存储池/kvm-vm，并将此存储池分配给虚拟机 new_vm 使用。

第 7 章

容器技术简介

信息技术的发展史就是一部“不断遇到新问题新挑战，然后解决问题应对挑战”的历史。云计算解决了信息系统基础设施中计算、存储、网络等几个方面的弹性问题，但是它还面临两个问题，这就是应用的扩展性和迁移性。容器技术就是一种解决方案。

本章要点

◎ 容器的定义
◎ 容器与虚拟机
◎ 容器的发展史
◎ 容器的标准化
◎ 容器的应用场景

7.1 容器的定义

容器是一个标准化的单元，是一种轻量级、可移植的软件打包技术。它将软件代码及其相关依赖文件打包，使应用程序可以在任何计算环境中运行。容器就像一个标准化的集装箱，能够容纳很多不同类型的货物，方便运输移动，如图 7-1 所示。

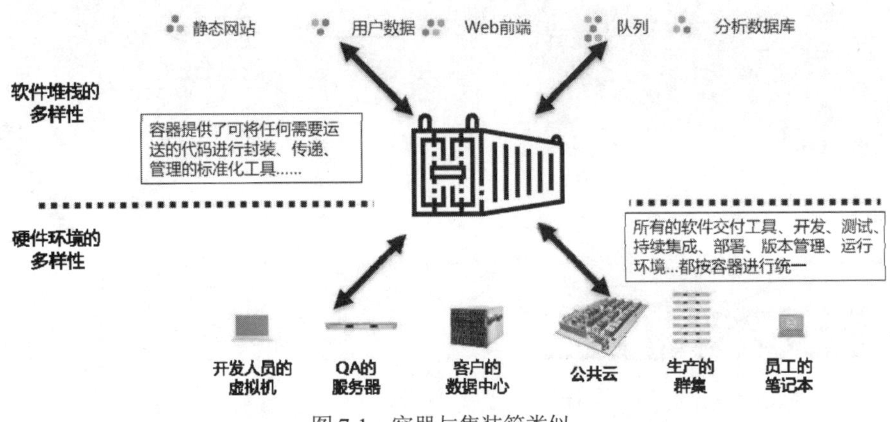

图 7-1　容器与集装箱类似

容器有以下特性：

(1) 软件打包：将软件打包成标准化单元以进行开发、迁移和部署。

(2) 隔离性：计算、存储、网络等资源彼此隔离。

(3) 高效性：轻量、快速启停、快速部署与迁移。

(4) 职责分工明确：开发人员专心写代码，运维人员专注基础环境配置。

7.2 实验环境部署

为了更好地体验容器的优势，我们将在CentOS Stream 9中安装著名的容器引擎——Docker。

安装 CentOS Stream 9 时，在 SOFTWARE SELECTION(软件选择)界面中，推荐选择"Minimal Install (最小化安装)"选项，以提高安装速度、系统性能及安全性，如图 7-2 所示。

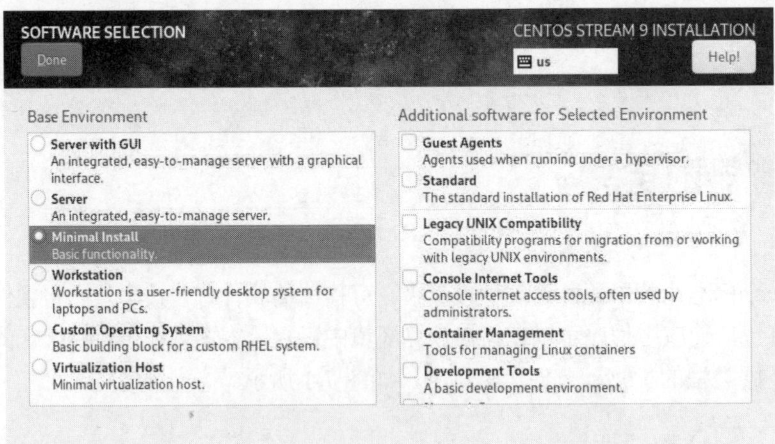

图 7-2　安装 CentOS Stream 9 时的软件安装选项

安装完成之后，通过 Putty 等 SSH 客户端软件登录主机，然后在命令行中安装 Docker 引擎，示例命令如下：

```
1 $ cat /etc/system-release
    CentOS Stream release 9

2 $ uname -a
    Linux docker1 5.14.0-191.el9.x86_64 #1 SMP PREEMPT_DYNAMIC Wed Nov 9 18:39:08 UTC 2022
      x86_64 x86_64 x86_64 GNU/Linux

3 $ dnf repolist
    repo id                 repo name
    appstream               CentOS Stream 9 - AppStream
    baseos                  CentOS Stream 9 - BaseOS
    extras-common           CentOS Stream 9 - Extras packages

4 $ sudo dnf -y install yum-utils
5 $ sudo dnf config-manager --add-repo \
https://download.docker.com/linux/centos/docker-ce.repo

6 $ dnf repolist
    repo id                 repo name
    appstream               CentOS Stream 9- AppStream
    baseos                  CentOS Stream 9 - BaseOS
    docker-ce-stable        Docker CE Stable - x86_64
    extras                  CentOS Stream 9 - Extras

7 $ sudo dnf -y install docker-ce docker-ce-cli containerd.io

8 $ sudo systemctl enable docker
9 $ sudo systemctl start docker
```

第 2 行命令的输出显示：Linux 内核的版本是 5.14.0。

第 3 行命令的输出显示当前环境中有 3 个软件仓库。其中：

- baseos 源提供了最小化系统所需要的一个包。
- appstream 源是应用程序流，也属于发行版本的一部分，包括额外的用户空间应用程序、运行时语言和数据库，以支持各种工作负载和用例。
- extras-common 源是额外的软件包。

第 4、第 5 行命令将创建新的 Docker 的软件仓库。

第 6 行命令输出显示新的仓库库名为 docker-ce-stable。

第 7 行命令安装 3 个 RPM 包，分别是 Docker 的服务器端、客户端和容器运行时(container runtime)。

安装完成之后，通过第 8 行命令将 Docker 服务器端软件设置为自动启动，第 9 行命令将其启动。

要快速验证实验环境是否配置正确，可以运行 Docker 官方提供的测试镜像 hello-world，示例命令如下：

```
$ docker run hello-world
  Unable to find image 'hello-world:latest' locally        #在宿主机中没有发现此镜像
  latest: Pulling from library/hello-world                 #从 Docker 官方仓库中下拉镜像
```

```
2db29710123e: Pull complete                                   #下拉的过程会有进度显示
Digest: sha256:faa03e786c97f07ef34423fccceeec2398ec8a5759259f94d99078f264e9d7af
Status: Downloaded newer image for hello-world:latest
# 下面是容器启动后自动执行命令的输出:
Hello from Docker!
This message shows that your installation appears to be working correctly.

To generate this message, Docker took the following steps:
 1. The Docker client contacted the Docker daemon.
 2. The Docker daemon pulled the "hello-world" image from the Docker Hub.
    (amd64)
 3. The Docker daemon created a new container from that image which runs the
    executable that produces the output you are currently reading.
 4. The Docker daemon streamed that output to the Docker client, which sent it
    to your terminal.

To try something more ambitious, you can run an Ubuntu container with:
 $ docker run -it ubuntu bash

Share images, automate workflows, and more with a free Docker ID:
 https://hub.docker.com/

For more examples and ideas, visit:
 https://docs.docker.com/get-started/
```

7.3 容器与虚拟机

谈到容器技术，就不得不把它与虚拟机技术进行对比，因为它们都为应用程序提供封装和隔离。

传统的虚拟化技术，例如VMware、KVM、Xen、Hyper-V，都需要创建完整的虚拟机，如图7-3左图所示。在这个类型的虚拟化主机中，如果运行3台虚拟机，就需要有3个Guest OS、3套二进制文件和库。虚拟机的优势是隔离性比较好，一个Guest OS不会影响另外一个Guest OS。但是缺点也很明显，即重复的文件比较多、资源占用大。

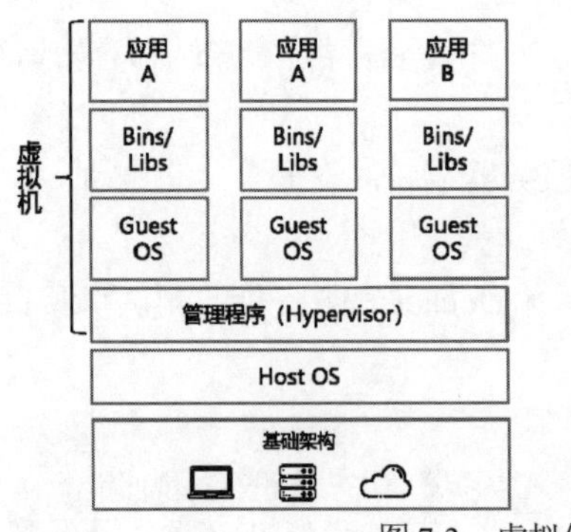

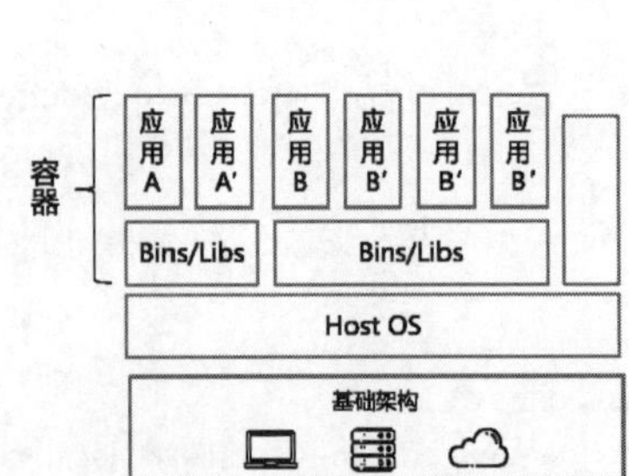

图7-3　虚拟化与容器对比

容器和虚拟机具有相似的资源隔离和分配优势，但功能不同，因为容器虚拟的是操作系统而不是硬件，容器更便携，更高效。容器没有虚拟化层，所以常被称为轻量级虚拟化技术，如图 7-3 右图所示。由于没有虚拟化层，运行在容器中的应用性能比运行在虚拟机中的更强。

容器的目标就是让应用程序的创建、部署和运行变得更加容易。容器由两部分组成：

- 应用程序本身。
- 应用程序依赖的资源，例如库文件或其他应用程序。

所有的容器都使用宿主机操作系统的内核空间的资源。下面通过示例考察这种共享宿主机操作系统内核的特性。

(1) 查看宿主操作系统的发行版本、内核的版本号，示例代码如下：

```
$ cat /etc/system-release
  CentOS Stream release 9
$ uname -a
  Linux docker1 5.14.0-191.el9.x86_64 #1 SMP PREEMPT_DYNAMIC Wed Nov 9 18:39:08 UTC 2022
  x86_64 x86_64 x86_64 GNU/Linux
```

(2) 运行 CentOS 容器，示例代码如下：

```
$ docker run -it centos bash
  Unable to find image 'centos:latest' locally
  latest: Pulling from library/centos
  a1d0c7532777: Pull complete
  Digest: sha256:a27fd8080b517143cbbbab9dfb7c8571c40d67d534bbdee55bd6c473f432b177
  Status: Downloaded newer image for centos:latest
# 下面是进入容器的操作，是一个 bash 的环境：
[root@9a3494cceb03 /]# cat /etc/system-release
CentOS Linux release 8.4.2105
[root@9a3494cceb03 /]# uname -a
Linux 9a3494cceb03 5.14.0-191.el9.x86_64 #1 SMP PREEMPT_DYNAMIC Wed Nov 9 18:39:08 UTC 2022
    x86_64 x86_64 x86_64 GNU/Linux
[root@9a3494cceb03 /]# exit
exit
```

docker 的 run 子命令可以执行一个容器，命令格式如下：

```
docker run [OPTIONS] IMAGE [COMMAND] [ARG...]
```

运行容器时，可以指定执行的一个命令，例如 bash shell 命令，其他两个选项的功能是：

- -i、-- interactive：即使未连接，也要保持 STDIN 打开。
- -t、--tty：分配一个伪 tty。

进入容器之后，可以发现发行版本是 CentOS Linux release 8.4.2105，内核版本是 5.14.0-191.el9.x86_64，内核版本与宿主机操作系统一致。

(3) 运行 Ubuntu 容器，示例代码如下：

```
$ docker run -it ubuntu bash
  Unable to find image 'ubuntu:latest' locally
  latest: Pulling from library/ubuntu
  e96e057aae67: Pull complete
  Digest: sha256:4b1d0c4a2d2aaf63b37111f34eb9fa89fa1bf53dd6e4ca954d47caebca4005c2
```

```
  Status: Downloaded newer image for ubuntu:latest
# 下面是进入容器的操作，是一个 bash 的环境：
root@e6f77651d494:/# cat /etc/os-release
PRETTY_NAME="Ubuntu 22.04.1 LTS"
NAME="Ubuntu"
VERSION_ID="22.04"
VERSION="22.04.1 LTS (Jammy Jellyfish)"
VERSION_CODENAME=jammy
ID=ubuntu
ID_LIKE=debian
HOME_URL="https://www.ubuntu.com/"
SUPPORT_URL="https://help.ubuntu.com/"
BUG_REPORT_URL="https://bugs.launchpad.net/ubuntu/"
PRIVACY_POLICY_URL="https://www.ubuntu.com/legal/terms-and-policies/privacy-policy"
UBUNTU_CODENAME=jammy
root@e6f77651d494:/# uname -a
Linux e6f77651d494 5.14.0-191.el9.x86_64 #1 SMP PREEMPT_DYNAMIC Wed Nov 9 18:39:08 UTC 2022
    x86_64 x86_64 x86_64 GNU/Linux
```

进入容器之后，可以发现发行版本是 Ubuntu 22.04.1 LTS，内核版本也是 5.14.0-191.el9.x86_64，内核版本与宿主机操作系统一致。

从以上示例可以看出：所有容器均共享宿主机操作系统的内核。由于启动容器不需要启动整个操作系统，所以容器部署和启动速度很快、开销更小，也更容易迁移。

容器和虚拟机之间的区别主要表现在操作系统支持、安全性、可移植性(便携性)、性能等 4 个方面，具体内容如表 7-1 所示。

1. 操作系统支持

每个虚拟机的 Guest 操作系统，都位于宿主机操作系统上，这使得虚拟机比较重。容器共享宿主机操作系统，所以是轻量级的，系统开销非常低，启动速度快。容器适用于在单个操作系统内核上运行多个应用程序的情况。但是，如果需要在不同操作系统上运行应用程序或服务器，就需要虚拟机，例如需要同时运行 Linux 应用程序和 Windows 应用程序。

2. 安全性

因为虚拟机不共享操作系统，所以有很强的隔离性。与容器相比，它们更安全。容器的风险点是用于共享主机操作系统的内核。

3. 便携性

因为容器没有单独的操作系统，所以很容易携带。一个容器可以移植到不同的操作系统中，并且可以立即启动。由于虚拟机有单独的操作系统，通常比较大，所以移植虚拟机要比容器难一些，时间长一些。对于必须在不同平台上进行开发和测试的应用程序，容器是理想的选择。

4. 性能

单纯比较虚拟机和容器的性能，其实不是很公平，因为它们有不同的适用场景。容器的轻量级体系结构，资源密集度较低，启动速度非常快，不需要将资源永久分配给容器，资源使用情况因其负载或流量而异。与虚拟机相比，扩展和复制容器也是一项简单的任务，因为不需要在其中安装操作系统。一台服务器上运行几十个虚拟机已经很不错了，但是运行容器，可以轻松达到上千个。

表 7-1　虚拟化与容器对比

虚拟机	容器
硬件级进程隔离	操作系统级进程隔离
每个虚拟机都有一个单独的操作系统	每个容器可以共享操作系统
在几分钟内启动	在几秒钟内启动
虚拟机文件通常比较大(GB 级)	容器文件通常比较小(KB/MB 级)
不太容易找到现成的虚拟机	很容易获得预先构建的容器
虚拟机可以轻松迁移到新主机	容器被销毁并重新创建而不是移动
创建虚拟机需要相对较长的时间	容器可以在几秒钟内被创建
资源使用多	资源使用少

从容器和虚拟机的使用场景来说，它们有着各自的应用场景，虽然会有一些重叠，两者的区别仍很明显。例如，虚拟机更适合比较重或庞大的单体应用和场景。对操作系统资源要求多时，使用虚拟机更为合适。而容器更适合轻量级的应用，迭代较多，如微服务，在服务器上运行更多的应用，适合在云环境快速迁移。另外，容器既可以运行在物理机之上，也可以运行在虚拟机中。

7.4 容器的发展史

了解容器技术的发展史有助于更好地理解容器技术，如图 7-4 所示。

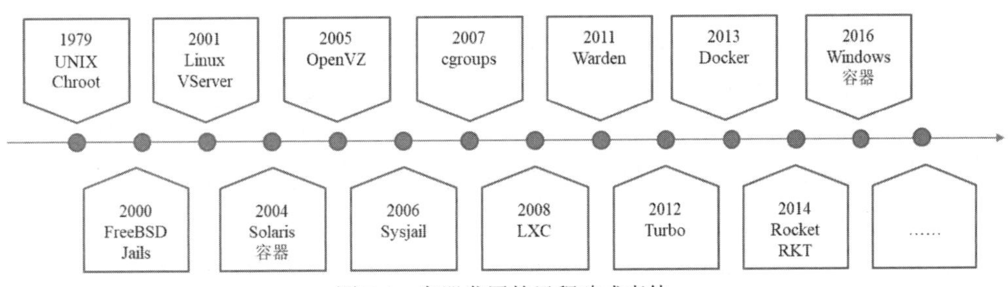

图 7-4　容器发展的里程碑式事件

容器技术最早可以追溯到 1979 年 UNIX 系统中的 chroot 技术，它是 change root 的缩写，最初是为了方便切换 root 目录，为每个进程提供了文件系统资源的隔离，这也是操作系统虚拟化思想的起源。

2000 年，FreeBSD 吸收并改进了 chroot 技术，发布了 Jails。除文件系统隔离外，还添加了用户和网络资源等的隔离，还能给每个 Jail 分配一个独立 IP，进行一些相对独立的软件安装和配置。

2001 年，SWsoft 公司发布了非开源的虚拟化软件 Virtuozzo(2005 年提供开源的版本)，提供了服务器整合、灾难恢复等功能。同一年，Linux VServer 发布，它延续了 FreeBSD Jails 的思想，在一个操作系统中隔离文件系统、CPU、网络和内存等资源，形成安全上下文(security context)。

2004 年，Sun 公司发布了 Solaris Containers。这是第一次使用“容器”这一术语。通过 Zone 提供的二进制隔离，Zone 是作为在操作系统实例内一个完全隔离的虚拟服务器而存在的。

2005 年，SWsoft 公司发布了开源的 OpenVZ。它是通过给 Linux 内核打补丁来提供虚拟化、隔离、资源管理和检查点功能，OpenVZ 的出现，标志着内核级别的虚拟化真正成为主流，之后不断有相关的技术被加入。

2006 年，DragonFly BSD 发布了 Sysjail。它最初的设计灵感来自 FreeBSD Jails，属于用户空间的虚拟机，只支持 OpenBSD、NetBSD 和 MirOS。

2006 年，Google 发布了 Process Containers。它记录和隔离每个进程的资源使用，包括 CPU、内存、硬盘 I/O、网络等。2007 年改名为 cgroups(control groups)，并被纳入到 2.6.24 版本的 Linux 内核中。

2008 年，出现了第一个比较完善的 LXC 容器技术。基于 Linux 内核中的 cgroups 和 namespaces 技术实现，不需要打补丁，就能运行在任意 vanilla 内核的 Linux 上。

提示

kernel.org 提供的就是 vanilla 内核，而不同的 Linux 发行版本可能会在这个原始内核基础上进行一些改动。例如，修改 bug，增加对新设备的支持等。

2011 年，Cloud Foundry 发布了 Warden。和 LXC 不同，它可以工作在任何操作系统上，除了提供守护进程之外，它还提供了管理容器的 API。

2012 年，Turbo.net 推出只支持 Windows 操作系统的非开源容器化软件。它是基于自有应用程序虚拟化引擎实现的容器。

2013 年，Docker 诞生。Docker 最早是 dotCloud 公司(Docker 公司的前身，是一家 PaaS 公司)的内部项目，Docker 最初使用 LXC，后来替换为 libcontainer。Docker 围绕容器构建了一套完整的生态，包括容器镜像标准、容器 Registry、REST API、CLI、容器集群管理工具 Docker Swarm 等，大获成功，现在 Docker 已基本成为容器技术的代名词。

提示

libcontainer 是用于容器管理的包，它基于 Go 语言实现，通过管理 namespaces、cgroups、capabilities 以及文件系统来进行容器控制。

2014 年 CoreOS 创建了 RKT。为了改进 Docker 在安全方面的缺陷，重写了一个容器引擎，相关容器工具产品包括服务发现工具 etcd 和网络工具 flannel 等。

2016 年，微软公司发布基于 Windows 的容器技术 Hyper-V Container，它的原理和 Linux 下的容器技术类似，可以保证在某个容器里运行的进程与外界是隔离的，且兼顾虚拟机的安全性和容器的轻量级。

2016 之后，进入容器的黄金时代，各种技术已经成熟，容器的安全性更好，管理工具也不更丰富，例如占有最高的容器管理平台 Kubernetes，自 2016 年被纳入云计算原生计算基金会(CNCF)以来，VMWare、Azure、AWS 甚至 Docker 都在其基础设施之上宣布了对它们的支持。

7.5 容器的标准化

在现阶段，Docker 几乎成了容器的代名词，甚至有些初学者错误地认为容器就是 Docker。就像关系型数据库有很多种一样，容器也有好多种，Docker 仅仅是其中比较流行的一种产品，其他的还有 RKT 和 CoreOS。

2013 年之后，容器技术百花齐放、争奇斗艳，这也产生了分歧，所以需要有一系列标准来规范它，不然很容易导致技术的碎片化、冲突和冗余。2015 年，由 Google、Docker、CoreOS、Red Hat、IBM、Microsoft 等厂商联合发起的开放容器倡议(Open Container Initiative，OCI)成立了。

OCI 是一个轻量级的、开放的治理结构，由 Linux 基金会赞助并管理，其主要任务之一是围绕容器格式和运行时创建开放的行业标准。镜像、容器与进程之间的关系如图 7-5 所示。OCI 目前包含运行时规范(Runtime Spec)和镜像规范(Image spec)两个规范。

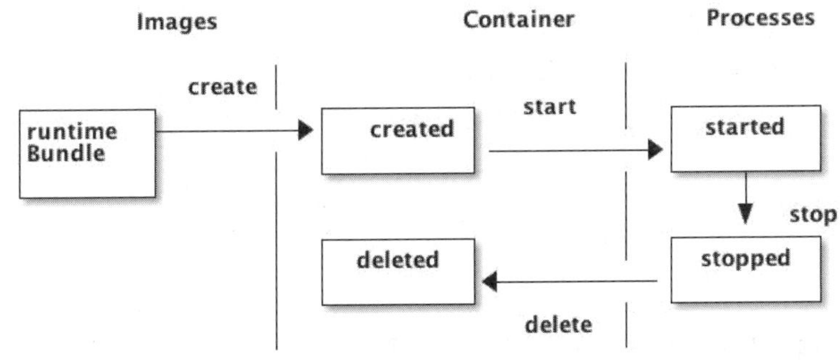

图 7-5　镜像、容器与进程之间的关系

镜像、容器与进程之间的关系如图 7-5 所示。容器运行时状态有：

(1) 运行时规范概述了如何运行容器。容器的状态包括：

- Creating：使用 create 命令创建容器，这是一个中间状态。
- Created：容器已经被创建，但是还没有运行，表示镜像和配置没有错误，容器能够在当前平台中运行。
- Running：容器处于运行状态，容器内进程处于 up 状态，正在执行用户设定的任务。
- Stopped：容器运行完成，或者运行出错。在此状态下，容器还有很多信息保存在平台中，并没有完全被删除。

(2) 容器镜像标准概述了镜像文件的组成，如图 7-6 所示。通常包括以下部分：

- 文件系统：以层(layer)形式保存的文件系统，每层保存了和上层之间变化的部分，涉及层应该保存哪些文件，如何表示增加、修改和删除文件等的信息。
- config 文件：保存了文件系统的层信息(如每层的散列值和历史信息)，以及容器运行时需要的一些信息(如环境变量、工作目录、命令参数、mount 列表等)，指定了镜像在某个特定平台和系统的配置。
- manifest 文件：镜像的 config 文件索引，存储与平台有关的信息。
- index 文件：可选的文件，指向不同平台的 manifest 文件，用于保证镜像可以跨平台使用。

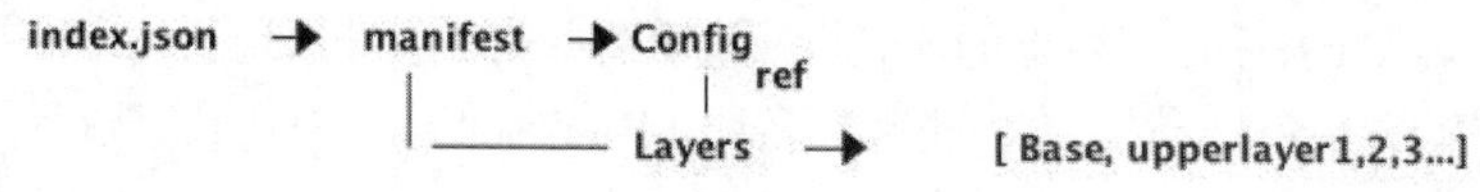

图 7-6　镜像文件的组成

7.6 容器的应用场景

在云计算时代，如果需要快速部署业务系统，就必须脱离底层物理硬件的限制；同时必须可以在“任何时间、任何地点”获取。因此，需要一种新型的部署应用程序的方式，实现快速分发和部署，而这正是容器技术的最大优势。

以传统的 LAMP 应用为例，传统的部署方式可能是这样的：

(1) 安装 Linux、Apache、MySQL 和 PHP 及运行所依赖的环境，还可能需要安装辅助工具，例如 phpMyadmin。

(2) 对每个组件进行配置，包括创建必要的用户、配置参数等。

(3) 进行功能测试。如果不正常，还需要进行调试追踪，这意味着需要更多的时间成本和不可控的风险。

最麻烦的事情是业务迁移，例如，从本地数据中心迁移到公共云，往往需要对每个组件进行重新部署和调试。这些琐碎而无趣的“体力活”，极大地降低了工作效率。究其根源，是由于这些应用直接部署在底层操作系统上，无法保证同一应用在不同的环境中行为一致。

而容器提供了一种“聪明”的方式，例如通过 Docker 打包应用、解耦应用和运行平台，如图 7-7 所示。这意味着迁移的时候，只需要在新的服务器上启动需要的容器就可以了，无论新旧服务器是否是同一类型的平台。这无疑将帮助我们节约大量的宝贵时间，并降低部署过程出现问题的风险。

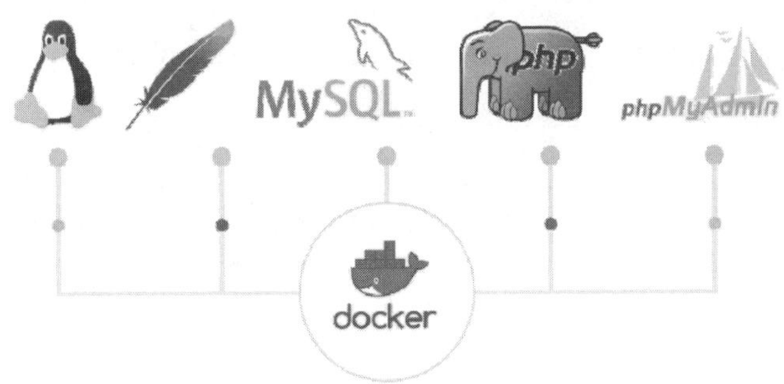

图 7-7　通过容器技术打包应用程序

容器的应用场景十分丰富，还是以 Docker 为例，一般有以下典型的应用场景。

(1) 简化配置。

Docker 提供了在多种基础设施、多种平台上，使用自己的配置运行应用程序的能力，而无需虚拟机的开销。Docker 允许将配置文件放入代码中，并通过不同的环境变量进行传递和部署。因此，相同的镜像可以在不同的环境中使用。这将使得基础架构需求与应用程序环境分离。

(2) 提高开发人员的生产力。

在开发环境中，通常我们既希望开发环境尽可能接近生产环境，又希望开发速度尽可能快，便于交互使用。为了实现第一个目标，需要开发环境能够反映生产应用程序的运行方式，Docker 的低开销，使得在通常内存容量较低的开发环境中，几十个服务可以在不同的容器中运行。为了实现第二个目标，可以利用 Docker 的共享卷，将宿主机操作系统上的应用程序代码直接用于容器。通过这种方式，开发人员可以从自己选择的平台和编辑器来编辑源代码，由于 Docker 容器与开发环境是一套源代码，所以可以立即看到更改后的效果。

(3) 服务器整合。

Docker 的应用程序隔离功能允许整合多台服务器以节省成本，而无需多个操作系统的内存占用，也无需跨实例共享未使用的内存。与虚拟机相比，Docker 提供了密度更大的服务器整合。

(4) 多租户技术。

传统应用程序要改造成多租户应用，需要进行大量的改写，成本很高。使用容器，给一个租户一个容器，容器之间相互隔离，从而避免了重写应用程序，所以仅需付出很低的成本就可以将传统应用程序改造成多租户的应用。

(5) 代码管道化管理。

配置的简化对代码管道化管理有很大的影响。在传统开发模式下，当代码从开发人员的计算机传输到生产环境时，需要经过许多不同的环境才能达到目的。在这一过程中，每一项都可能有细微的差异。Docker 为应用程序从开发到生产提供了一个一致的环境，简化了代码开发和部署流程。Docker 镜像的不变特性，帮助我们在整个开发过程中，实现应用程序运行时环境的零更改。

(6) 应用程序隔离。

不同应用程序可能对底层组件的版本配置有不同要求，例如有两个 REST API 应用程序都需要 Apache Web 服务器，但是使用的 Apache 版本和依赖项不同。在 Docker 中，可以为不同的容器配置不同的 Apache 版本及组件，这个场景也被为“依赖隔离”。

(7) 调试功能。

Docker 提供了许多工具用于对应用进行调试，例如检查容器和容器版本的能力，以及区分两个容器的能力，这在修复应用程序时非常有用。

(8) 快速部署。

可以在很短时间内创建和更新 Docker 容器。通过容器部署应用程序，不需要修改宿主机操作系统，而是像运行普通的应用程序一样。此外，Docker 镜像的不变特性，让这些应用总是按照它们一直在工作的方式和应该的方式来运行。

7.7 习题

1. [单选题]Docker 是(　　)。

   A. 虚拟机　　B. 半虚拟化技术

   C. 全虚拟化技术　　D. 开源的应用容器引擎

2. [单选题]Docker 与 KVM 虚拟化技术的区别是(　　)。

   A. Docker 容器启动快，资源占用小，操作系统级虚拟化技术

   B. KVM 虚拟化启动快，资源占用小，操作系统级虚拟化技术

   C. 需要根据具体的业务负载来确定区别

   D. 没区别

3. [单选题]Docker 与传统虚拟机的区别是(　　)。

   A. Docker 容器的启动速度是秒级，而传统虚拟机是分钟级

   B. Docker 容器在计算能力损耗上接近 50%，而传统虚拟机几乎无损耗

   C. Docker 容器单机可启动几十台，而传统虚拟机可以启动上千台

   D. Docker 容器在隔离性上是完全隔离，而传统虚拟机则采用资源限制

4. [单选题]Docker 所依赖的虚拟化技术为(　　)。

   A. 完全虚拟化　　B. 准虚拟化

   C. 操作系统层虚拟化　　D. 桌面虚拟化

5. [单选题]Docker 容器的实质是(　　)。

   A. 线程　　B. 进程

   C. 程序　　D. 文件

第 8 章

Docker镜像管理

镜像是容器的基石，容器是镜像的运行实例，有了镜像才能启动容器。本章将学习镜像的作用、内部结构、管理、分发和构建等方面的知识。

本章要点

- ◎ 镜像的作用
- ◎ 镜像的结构
- ◎ 查看镜像信息
- ◎ 搜索镜像
- ◎ 删除和清理镜像
- ◎ 创建镜像
- ◎ 保存和加载镜像
- ◎ 集中管理镜像

8.1 镜像的作用

Docker 采用的是客户端/服务器架构。Docker 客户端程序和守护程序可以使用 REST API、UNIX 套接字或网络接口进行通信，客户端发送请求指令，守护程序负责构建、运行容器的工作。Docker 客户端和守护程序既可以在同一主机上运行，也可以用 Docker 客户端连接到远程的 Docker 守护程序，如图 8-1 所示。

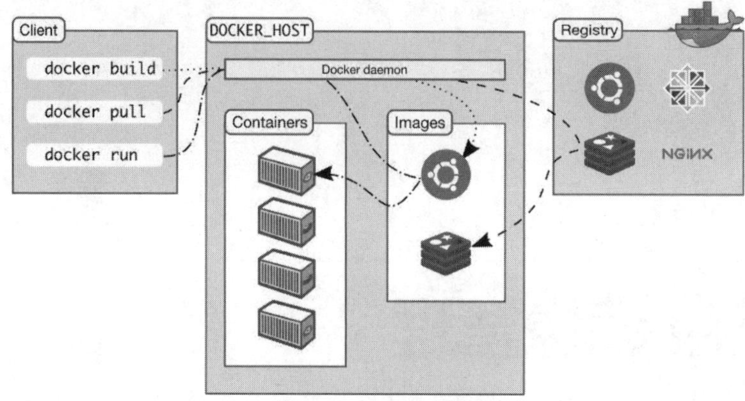

图 8-1　Docker 架构

(1) Docker 客户端程序(Client)，例如 docker 命令，将请求指令发送给守护程序。

(2) Docker 守护程序(Docker daemon)侦听 Docker API 请求并管理 Docker 资源，例如镜像、容器、网络和卷。

(3) Docker 注册表(Registry)是公共或私有的镜像仓库，例如 Docker Hub(hub.docker.com)就是目前世界上最大的公共镜像仓库，我们也可以构建自己的私有注册表。Docker 默认使用 Docker Hub。

镜像是一个只读模板，包含用于创建容器的说明与数据。既可以从无到有地创建全新的镜像，也可以基于一个镜像再创建新的镜像，并带有一些有自己特色的、额外的属性和数据。例如，在一个 Ubuntu 的镜像的基础上安装 Nginx 等软件包，构建提供 Web 服务的新镜像。

8.2 获取镜像

拥有镜像是运行容器的前提。如果运行容器时，宿主机没有所需的镜像文件，就会自动从仓库中下拉镜像。示例命令如下：

```
1 $ docker run -i -t ubuntu bash
    Unable to find image 'ubuntu:latest' locally
    latest: Pulling from library/ubuntu
    e96e057aae67: Pull complete
    Digest: sha256:4b1d0c4a2da89fa1bf53dd6e4ca954d47caebca4005c2********
    Status: Downloaded newer image for ubuntu:latest
    root@4e45e82143ca:/# exit
    exit

2 $ docker run -i -t ubuntu bash
    root@fdfdd63519bc:/# exit
    exit

3 $ docker image ls
    REPOSITORY     TAG          IMAGE ID        CREATED              SIZE
```

```
ubuntu          latest          a8780b506fa4     2 weeks ago          77.8MB
```

第 1 行命令的输出显示：由于宿主机上没有 Ubuntu 容器的镜像，所以会先从仓库中下载镜像，然后再运行。下载过程中会显示每一层的 ID 的前 12 位，下载结束后会显示镜像完整的 SHA-256 的摘要，以确保下载一致性。

第 2 行命令再次运行，由于宿主机拥有所需镜像，所以可以直接运行。

第 3 行命令查看宿主机上的所有镜像信息，显示的信息包括：镜像的存储库名称、标签、短 ID、创建时间、大小等。

提示

容器和镜像的 ID 是 64 个字符的 SHA-256 ID，通常被称为“长 ID”。为了提高效率，可以只使用前 12 个字符，通常被称为“短 ID”。在保证唯一性的前提下，也可以只用前 4 个字符来表示容器或镜像。

除了在运行时自动下载镜像之外，还可以通过 pull 子命令手工下载。通过以下命令查看它的联机帮助：

```
4 $ docker pull --help
    Usage: docker pull [OPTIONS] NAME[:TAG|@DIGEST]
    Pull an image or a repository from a registry
    Options:
      -a, --all-tags                  Download all tagged images in the repository
          --disable-content-trust     Skip image verification (default true)
          --platform string           Set platform if server is multi-platform capable
      -q, --quiet                     Suppress verbose output
```

NAME 是镜像的名称，用来区分镜像。TAG 是可选的标签，常被用于标识镜像的版本信息。TAG 的默认值为 latest。手工下载最新版本的 Ubuntu 镜像，示例命令如下：

```
5 $ docker pull ubuntu
    Using default tag: latest
    latest: Pulling from library/ubuntu
    e96e057aae67: Downloading [=========================>          ]  18.38MB/30.43MB
```

等下载完毕，会有如下提示：

```
    e96e057aae67: Pull complete
    Digest: sha256:4b1d0c4a2d2aaf63b37111f34eb9fa89fa1bf53dd6e4ca954d****
    Status: Downloaded newer image for ubuntu:latest
    docker.io/library/ubuntu:latest
```

下拉指定版本的 Ubuntu 镜像，示例命令如下：

```
6 $ docker pull ubuntu:18.04
    18.04: Pulling from library/ubuntu
    a404e5416296: Downloading [==>                                 ]  1.67MB/26.71MB
```

等下载完毕，会有如下提示：

```
    a404e5416296: Pull complete
```

```
    Digest: sha256:ca70a834041dd1bf16cc38dfcd24f0888ec4fa431e09f3344f****
    Status: Downloaded newer image for ubuntu:18.04
    docker.io/library/ubuntu:18.04

7 $ docker image ls
    REPOSITORY      TAG      IMAGE ID       CREATED       SIZE
    ubuntu          latest   a8780b506fa4   2 weeks ago   77.8MB
    ubuntu          18.04    71eaf13299f4   3 weeks ago   63.1MB
```

第 7 行命令的输出显示：宿主机现在有两个版本的 Ubuntu 镜像。

提示

在生产环境中，建议使用 NAME+TAG 的格式明确指定版本，避免版本之间的差异，以增加确定性。

下载的镜像是压缩包，会自动解压缩到/var/lib/docker 子目录中。

8.3 镜像的结构

Docker 镜像是由一个或多个只读层(read-only layer)堆叠在一起形成的统一视图。容器又在其上叠加了一个可读、可写的层，如图 8-2 所示。

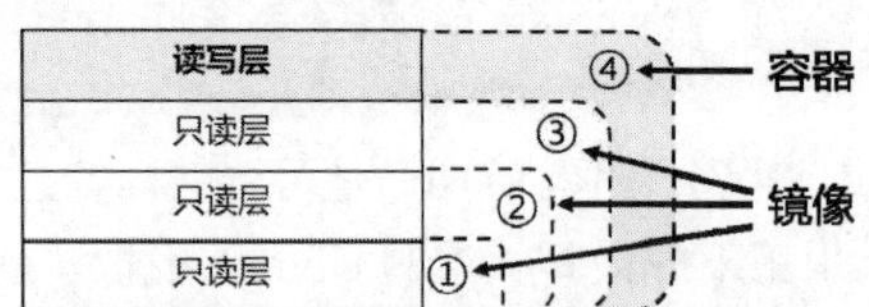

图 8-2　Docker 镜像与容器的层叠结构

站在镜像的角度来看(图 8-2 中的第 3 层)，镜像是由多个只读层叠加在一起形成的。这种技术被称为统一文件系统(union file system)，对外提供一个统一的视角，从而隐藏了多层的存在，对于最终用户来说是透明的，只看到最上面一层，如图 8-3 所示。

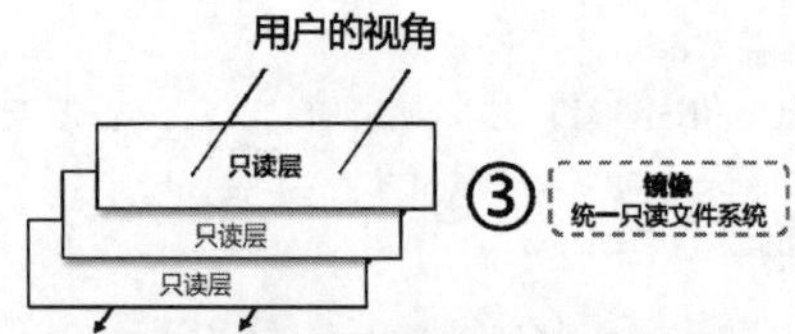

图 8-3　Docker 镜像的统一文件系统

在镜像中，除了最底层之外，其他每一层都有一个指向父层的指针。当下载镜像时，会根据指针信息将各层都下载下来，示例命令如下：

```
$ docker pull httpd
  Using default tag: latest
```

```
latest: Pulling from library/httpd
61994089e28e: Pull complete
8de88a91bde5: Pull complete
05820377a11a: Pull complete
70618b6c8070: Pull complete
70618b18b6e8: Pull complete
Digest: sha256:2d1f8839d6127e400ac5f65481d8a0f17ac46a3b91de40b01e6******
Status: Downloaded newer image for httpd:latest
docker.io/library/httpd:latest
```

从 pull 子命令的输出中可以看出这个镜像的层次结构如图 8-4 所示。

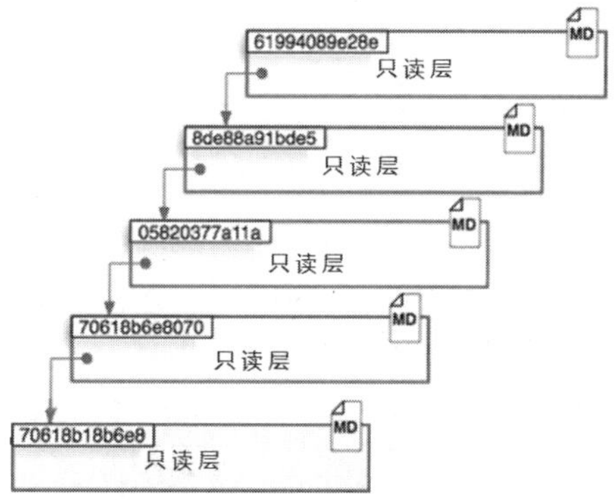

图 8-4　Docker 镜像层之间的关系

每一层的结构如图 8-5 所示，主要包括以下元素：

- 层 ID：镜像层的标识。
- 元数据(metadata)：镜像层的属性信息。
- 指针：指向父层 ID 的标识。
- 文件数据：与父层之间的差异数据。

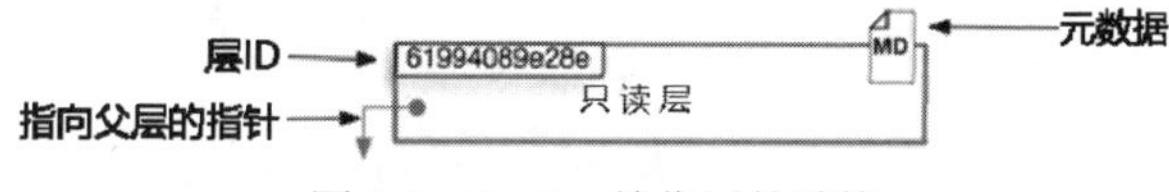

图 8-5　Docker 镜像层的结构

8.4 Docker 的存储驱动程序

Docker 可以在 Linux、Windows 等操作系统上运行。针对不同的操作系统，Docker 使用灵活的、可插拔的存储驱动程序来支持宿主机的文件系统。目前 Docker 在 Linux 上提供的存储驱动程序有：

- overlay
- overlay 2
- fuse-overlayfs

- btrfs 和 zfs
- vfs
- aufs
- devicemapper

推荐在 Linux 中使用 Overlay 2 存储驱动程序。查看存储驱动程序的版本信息，示例命令如下：

```
1 $ cat /etc/system-release
    CentOS Stream release 9

2 $ docker info
    Client:
     Context: default
     Debug Mode: false
     Plugins:
      app: Docker App (Docker Inc., v0.9.1-beta3)
      buildx: Docker Buildx (Docker Inc., v0.9.1-docker)
      scan: Docker Scan (Docker Inc., v0.21.0)

    Server:
     Containers: 0
      Running: 0
      Paused: 0
      Stopped: 0
     Images: 3
     Server Version: 20.10.21
     Storage Driver: overlay2
      Backing Filesystem: xfs
      Supports d_type: true
      Native Overlay Diff: true
      userxattr: false
……略……
```

Overlay 2 工作原理与 Overlay 类似，都是使用两个或更多个目录创建一个联合的目录，根据规则融合出目录及文件，构建一个统一的文件系统。此文件系统中所有的下层目录都是只读的，只有最高一层的目录是可读可写的。

下面通过手工创建一个简化版的 Overlay 文件系统，考察统一文件系统的特性。示例命令如下：

```
1 $ cd /tmp && mkdir overlay-example && cd overlay-example

2 $ mkdir container layer-1 layer-2 layer-3 layer-4 workdir
3 $ echo 111 >layer-1/file1
4 $ echo 222 >layer-2/file2
5 $ echo 333 >layer-3/file3

6 $ tree
    .
    ├── container
```

```
├── layer-1
│   └── file1
├── layer-2
│   └── file2
├── layer-3
│   └── file3
├── layer-4
└── workdir

7 $ sudo mount -t overlay overlay-example \
  -olowerdir=/tmp/overlay-example/layer-1:/tmp/overlay-example/layer-2:/tmp/overlay-example/layer-3,\
  upperdir=/tmp/overlay-example/layer-4,\
  workdir=/tmp/overlay-example/workdir \
  /tmp/overlay-example/container

8 $ sudo tree
.
├── container
│   ├── file1
│   ├── file2
│   └── file3
├── layer-1
│   └── file1
├── layer-2
│   └── file2
├── layer-3
│   └── file3
├── layer-4
└── workdir
    └── work
```

第 1 行的命令在/tmp 目录中创建一个名为 overlay-example 的子目录，并且进入到此目录中。

第 2 行的命令再创建 6 个子目录。可以将 container 子目录想象成容器的统一视图，是一个挂载(mount)目录，将 layer-1、layer-2、layer-3 子目录想象成镜像层，将 layer-4 子目录想象成容器的可读可写层，workdir 子目录是 Overlay 文件系统的临时目录。

第 3~第 5 行的命令在 3 个下层目录中创建测试文件。

第 6 行的 tree 命令查看目录结构与文件。

第 7 行的 mount 命令挂载一个 overlay 文件系统，挂载目录为/tmp/overlay-example/ container。其中 overlay-example 表示挂载的源，-o 后面选项有(-o 与选项之间不能有空格)：

- lowerdir：指定 3 个下层目录，可以将它们理解为只读的镜像层。
- upperdir：指定上层目录，可以将它理解为可读可写的容器层。
- workdir：指定临时的工作目录。
- /tmp/overlay-example/container：文件系统的挂载点，是用户访问的入口。

从第 8 行命令的输出可以看出：mount 命令合并出的目录是 container，其中的 file1、file2 和 file3 等 3 个文件，都来自下层的目录，如图 8-6 所示。当用户读取这些文件时，文件系统会自上而下进行查找，例如：读取 file1，如果 layer-4 没有，就找 layer-3；如果 layer-3 也没有，就找 layer-2，以此类推，直到在 layer-1 中找到。这个过程是由 Overlay 文件系统自动完成的，对用户来说是透明的。

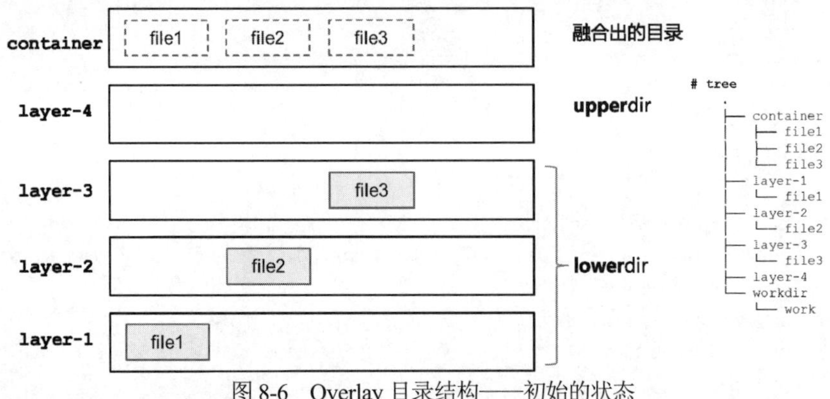

图 8-6　Overlay 目录结构——初始的状态

在初始状态下，upperdir 目录(layer-4)是空的。如果向 container 中写入一个新的文件 new-file，它会被保存在 layer-4 子目录中，如图 8-7 所示。

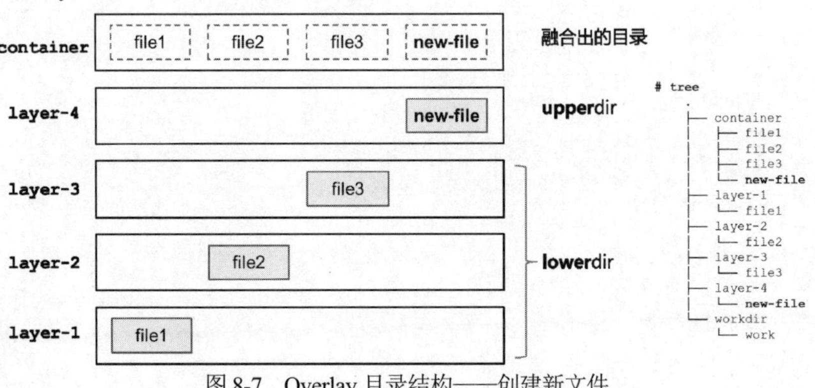

图 8-7　Overlay 目录结构——创建新文件

如果修改了文件 file3，由于 Overlay 采用的是 Copy on Write 机制，所以会先将文件 file3 复制到 layer-4 子目录中，然后再进行修改保存。layer-4 子目录中始终保存着文件的最新版本，下层的目录存储的是最初的原始文件，不会发生变化，如图 8-8 所示。

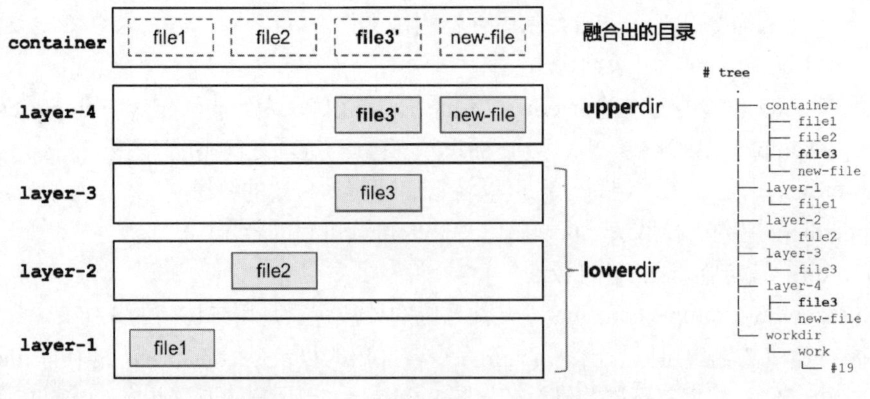

图 8-8　Overlay 目录结构——修改文件

如果删除文件 file2，layer-2 子目录中的原始文件 file2 并没有发生变化，而是在 layer-4 子目录中创建带有一个特殊属性标记的文件，表示 file2 已经被删除，如图 8-9 所示。这个特殊属性标记是 Character file 属性，Overlay 文件系统把它被称为 whiteout(短暂失明)属性，用于表示被删除的文件。示例命令如下所示：

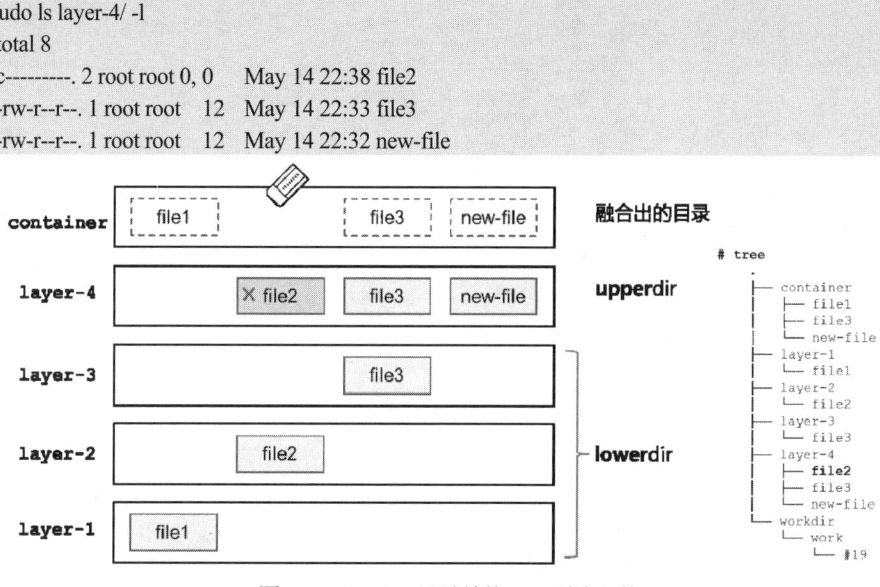

```
$ sudo ls layer-4/ -l
  total 8
  c---------. 2 root root 0, 0    May 14 22:38 file2
  -rw-r--r--. 1 root root    12   May 14 22:33 file3
  -rw-r--r--. 1 root root    12   May 14 22:32 new-file
```


图 8-9　Overlay 目录结构——删除文件

通过这个实验，我们考察了 Overlay 文件系统的基本原理，而 Docker 真实的情况要比这复杂很多，例如：层的标识、metadata、指针等。通过 Overlay 文件系统最后达到的效果是：容器=镜像+读写层。镜像是只读的，类似实验中下层的子目录(layer-1、layer-2、layer-3)。在第一次启动容器时，会在镜像层的基础上添加一个读写层(layer-4)，用户对容器的所有写操作都保存在这一层。一旦容器被删除，这个读写层也随之被销毁。

如果是 Linux 版的 Docker，镜像文件默认的存储目录是/var/lib/docker/overlay2/。下面通过一个简单的对照实验来考察存储结构，示例命令如下：

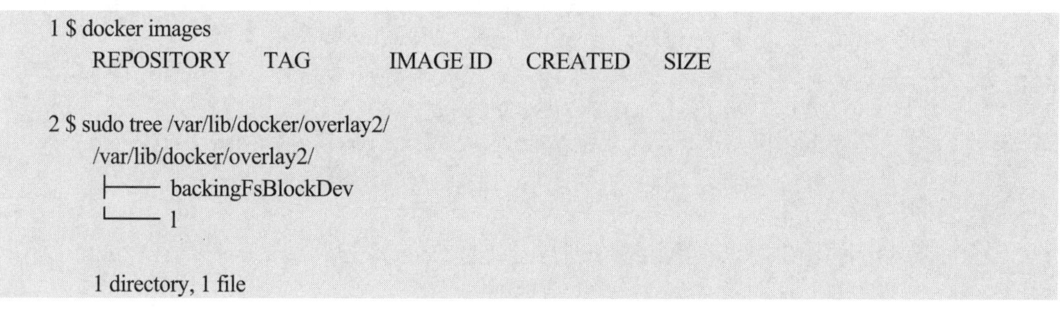

```
1 $ docker images
    REPOSITORY    TAG        IMAGE ID    CREATED    SIZE

2 $ sudo tree /var/lib/docker/overlay2/
    /var/lib/docker/overlay2/
    ├── backingFsBlockDev
    └── 1

    1 directory, 1 file
```

在没有镜像之前，这个目录是空的目录结构，并没有文件。

```
3   $ docker pull centos
    Using default tag: latest
    latest: Pulling from library/centos
    a1d0c7532777: Pull complete
    Digest: sha256:a27fd8080b517143cbbbab9dfb7c8571c40d67d534bbdee55bd6c473f432b177
```

```
    Status: Downloaded newer image for centos:latest
    docker.io/library/centos:latest

4   $ sudo ls -l /var/lib/docker/overlay2/
    total 0
    drwx--x---. 3 root root      30   Nov 18    12:29
    3eedcbb98ffbe71fc5b7daa  0339cb696a5d0293fafe8a9516ac72893c20446a7
    brw-------. 1 root root 253,  0   Nov 18    11:51 backingFsBlockDev
    drwx------. 2 root root      40   Nov 18    12:291
```

从第 3 行、第 4 行命令的输出可以看出：由于镜像文件只有一层，所以在目录/var/lib/docker/overlay2/中会新增一个子目录。

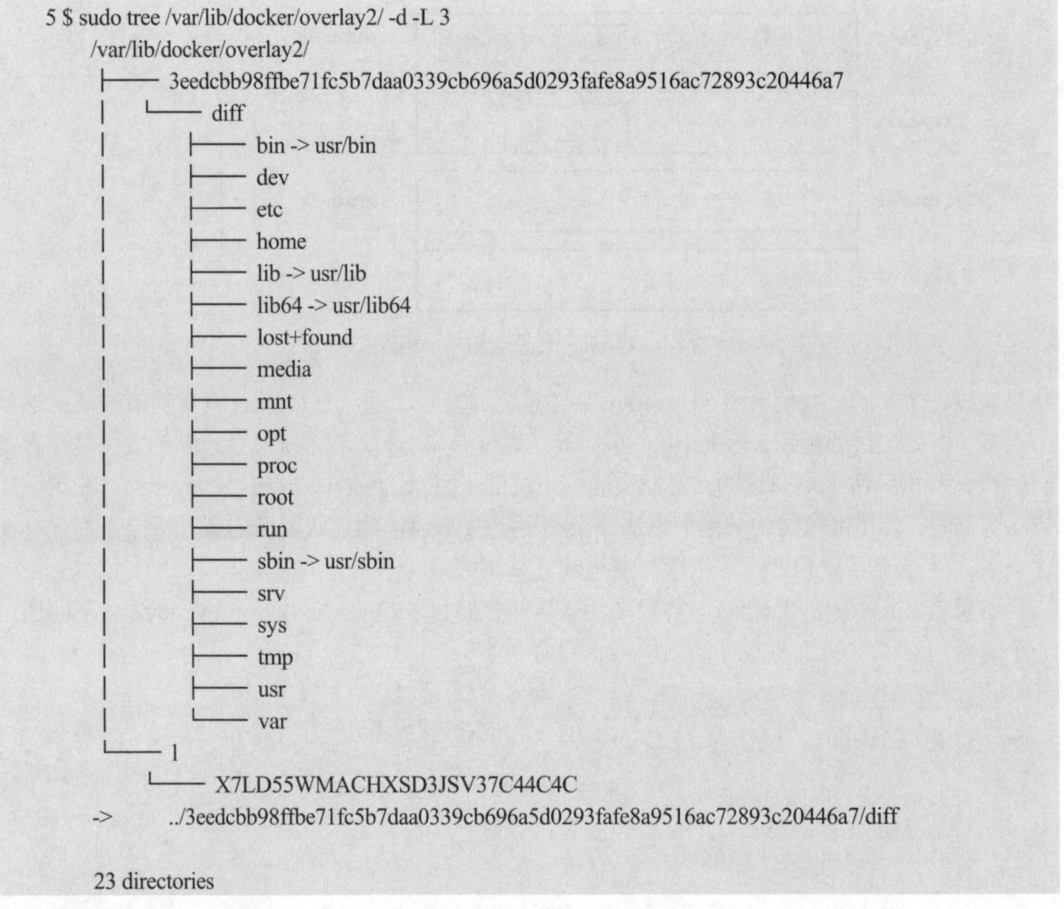

```
5 $ sudo tree /var/lib/docker/overlay2/ -d -L 3
    /var/lib/docker/overlay2/
    ├── 3eedcbb98ffbe71fc5b7daa0339cb696a5d0293fafe8a9516ac72893c20446a7
    │   └── diff
    │       ├── bin -> usr/bin
    │       ├── dev
    │       ├── etc
    │       ├── home
    │       ├── lib -> usr/lib
    │       ├── lib64 -> usr/lib64
    │       ├── lost+found
    │       ├── media
    │       ├── mnt
    │       ├── opt
    │       ├── proc
    │       ├── root
    │       ├── run
    │       ├── sbin -> usr/sbin
    │       ├── srv
    │       ├── sys
    │       ├── tmp
    │       ├── usr
    │       └── var
    └── 1
        └── X7LD55WMACHXSD3JSV37C44C4C
    ->      ../3eedcbb98ffbe71fc5b7daa0339cb696a5d0293fafe8a9516ac72893c20446a7/diff

    23 directories
```

从第 5 行命令的输出可以看出：这个新目录就是 CentOS 发行版本的目录结构。

8.5 查看镜像信息

查看 Docker 镜像信息的子命令如下：

- images/image ls：列出镜像。
- tag：为镜像添加标签。
- inspect：查看详细信息。
- history：查看镜像历史。

8.5.1 使用 images/image ls 子命令列出镜像

使用 images 和 image ls 子命令都可以列出宿主机中已有镜像的基本信息。它们有什么区别呢？查看它们的联机帮助，示例命令如下：

```
$ docker help images
  Usage: docker images [OPTIONS] [REPOSITORY[:TAG]]
  List images
  Options:
    -a, --all              Show all images (default hides intermediate images)
        --digests          Show digests
    -f, --filter filter    Filter output based on conditions provided
        --format string    Pretty-print images using a Go template
        --no-trunc         Don't truncate output
    -q, --quiet            Only show image IDs

$ docker help image
  Usage: docker image COMMAND
  Manage images
  Commands:
    build      Build an image from a Dockerfile
    history    Show the history of an image
    import     Import the contents from a tarball to create a filesystem image
    inspect    Display detailed information on one or more images
    load       Load an image from a tar archive or STDIN
    ls         List images
    prune      Remove unused images
    pull       Pull an image or a repository from a registry
    push       Push an image or a repository to a registry
    rm         Remove one or more images
    save       Save one or more images to a tar archive (streamed to STDOUT by default)
    tag        Create a tag TARGET_IMAGE that refers to SOURCE_IMAGE
  Run 'docker image COMMAND --help' for more information on a command.
```

images 子命令只能列出镜像的基本信息，而 imagels 子命令却是管理镜像子命令的合集，拥有多个子命令。单纯从获得镜像列表功能来看，它们是没有区别的，示例命令如下：

```
$ docker images
REPOSITORY  TAG      IMAGE ID       CREATED        SIZE
ubuntu      latest   a8780b506fa4   2 weeks ago    77.8MB
ubuntu      18.04    71eaf13299f4   3 weeks ago    63.1MB
centos      latest   5d0da3dc9764   14 months ago  231MB
```

命令的输出包含以下信息：

- REPOSITORY：存储库名称，例如，ubuntu 表示 ubuntu 系列的基础镜像。

- TAG：镜像的标签信息，通常是版本信息，用于标记来自同一个仓库的不同镜像的版本，还可以通过 tag 子命令为镜像创建新的标签。
- IMAGE ID：镜像的短 ID。如果镜像的 ID 相同，说明它们是同一个镜像。
- CREATED：镜像最后的更新时间。
- SIZE：镜像大小。这是该镜像的逻辑大小，由于相同的镜像层只会存储一份，所以物理上占用的存储空间会小于各镜像逻辑大小之和。

images 和 image ls 子命令还支持如下选项：

- -a,--all=true|false：列出所有(包括临时文件)镜像文件，默认为 false。
- --digests=true|false：列出镜像的数字摘要值，默认为 false。
- -f, --filter=[]：过滤列出的镜像，例如，dangling=true 表示只显示虚悬的镜像。
- --format="TEMPLATE"：控制输出格式，如.ID 表示 ID 信息，.Repository 代表仓库信息等。
- --no-trunc=true|false：截断输出结果中太长的部分，例如，镜像的 ID。默认为 true。
- -q, --quiet=true|false：仅输出 ID 信息，默认为 false。

提示

除使用 man 帮助之外，还可以在 https://docs.docker.com/engine/reference/run/中查看 Docker 命令行参考手册。

8.5.2 使用 tag 子命令为镜像添加标签

为了方便镜像管理，可以使用 tag 子命令为一个镜像添加多个 tag，它的语法结构如下：

```
$ docker tag --help
Usage: docker tag SOURCE_IMAGE[:TAG] TARGET_IMAGE[:TAG]
Create a tag TARGET_IMAGE that refers to SOURCE_IMAGE
```

名称：由斜杠分隔的名称。

TAG：只支持大小写字母、数字、下划线、句点和破折号。

给镜像 ubuntu:latest 添加新的标签，示例命令如下：

```
1 $ docker images
    REPOSITORY   TAG       IMAGE ID       CREATED          SIZE
    ubuntu       latest    a8780b506fa4   2 weeks ago      77.8MB
    ubuntu       18.04     71eaf13299f4   3 weeks ago      63.1MB
    centos       latest    5d0da3dc9764   14 months ago    231MB

2 $ docker tag ubuntu:latest myubuntu:0.9

3 $ docker images
    REPOSITORY   TAG       IMAGE ID       CREATED          SIZE
    myubuntu     0.9       a8780b506fa4   2 weeks ago      77.8MB
    ubuntu       latest    a8780b506fa4   2 weeks ago      77.8MB
    ubuntu       18.04     71eaf13299f4   3 weeks ago      63.1MB
```

```
centos              latest         5d0da3dc9764    14 months ago       231MB
```

从第 3 行命令的输出可以看出：镜像 myubuntu:0.9 的 ID 是 a8780b506fa4，与源镜像 ubuntu:latest 是一样的。这说明它们实际上是同一个镜像文件，只是名称与标签不同而已。通过 docker tag 命令添加的标签，与 Linux 文件系统的硬连接的作用类似。根据需要还可以采用多种引用方式。

(1) 通过引用 ID 创建新标签，示例命令如下：

```
$ docker tag 0e5574283393 fedora/httpd:version1.0
```

(2) 通过引用名称创建新标签，示例命令如下：

```
$ docker tag httpd fedora/httpd:version1.0
```

(3) 通过引用名称和标签创建新标签，示例命令如下：

```
$ docker tag httpd:test fedora/httpd:version1.0.test
```

(4) 为私有存储库创建新标签，示例命令如下：

```
$ docker tag 0e5574283393 myregistryhost:5000/fedora/httpd:version1.0
```

Docker 不支持删除标签和更改标签，没有类似 untag、renametag 之类的子命令，只能通过删除特定标签的镜像来实现，示例命令如下：

```
1 $ docker rmi --help
    Usage:   docker rmi [OPTIONS] IMAGE [IMAGE...]
    Remove one or more images
    Options:
      -f, --force          Force removal of the image
          --no-prune       Do not delete untagged parents

2 $ docker rmi myubuntu:0.9
    Untagged: myubuntu:0.9

3 $ docker images
    REPOSITORY          TAG            IMAGE ID        CREATED           SIZE
    ubuntu              latest         a8780b506fa4    2 weeks ago       77.8MB
    ubuntu              18.04          71eaf13299f4    3 weeks ago       63.1MB
    centos              latest         5d0da3dc9764    14 months ago     231MB

4 $ docker rmi ubuntu:latest
    Untagged: ubuntu:latest
    Untagged: ubuntu@sha256:4b1d0c4a2d2aaf63b37111f34eb9fa89fa1bf53dd6e4ca954d47caebca4005c2
    Deleted: sha256:a8780b506fa4eeb1d0779a3c92c8d5d3e6a656c758135f62826768da458b5235
    Deleted: sha256:f4a670ac65b68f8757aea863ac0de19e627c0ea57165abad8094eae512ca7dad

5 $ docker images
    REPOSITORY          TAG            IMAGE ID        CREATED           SIZE
    ubuntu              18.04          71eaf13299f4    3 weeks ago       63.1MB
    centos              latest         5d0da3dc9764    14 months ago     231MB
```

由于有两个标签指向同一个镜像，所以第 2 行的 rmi 子命令的输出中仅显示了 Untagged，这与删除文件系统的硬连接类似，并没有真正删除镜像，删除的是指针。

第 4 行的 rmi 子命令再删除另外一个镜像，从输出可以看出：也是先进行 Untag 操作，由于这是最后一个标签，所以还会真正地删除镜像。由于这个镜像文件由两层组成，所以进行了两次删除操作。

8.5.3 使用 inspect 子命令查看详细信息

使用 inspect 子命令可以获取该镜像的详细信息，包括作者、适用的架构、各层的数字摘要、对外提供的端口、环境变量、自动执行的命令等信息，示例命令如下：

```
1 $ docker inspect --help
    Usage:    docker inspect [OPTIONS] NAME|ID [NAME|ID...]
    Return low-level information on Docker objects
    Options:
      -f, --format string    Format the output using the given Go template
      -s, --size             Display total file sizes if the type is container
      --type string          Return JSON for specified type

2 $ docker inspect centos:latest
    [
        {
            "Id": "sha256:5d0da3dc976460b72c77d94c8a1ad043720b0416bfc16c52c45d4847e53fadb6",
            "RepoTags": [
                "centos:latest"
            ],
            "RepoDigests": [
              "centos@sha256:a27fd8080b517143cbbbab9dfb7c8571c40d67d534bbdee55bd6c473f432b177"
            ],
            "Parent": "",
            "Comment": "",
            "Created": "2021-09-15T18:20:05.184694267Z",
            "Container": "9bf8a9e2ddff4c0d76a587c40239679f29c863a967f23abf7a5babb6c2121bf1",
            "ContainerConfig": {
                "Hostname": "9bf8a9e2ddff",
                "Domainname": "",
                "User": "",
……略……

3 $ docker inspect -f {{".Architecture"}} centos
    amd64
```

inspect 子命令的返回结果是一个 JSON 格式的消息。如果只需要其中某一项信息，可以使用-f 选项指定格式，例如，获取镜像的 Architecture。

8.5.4 使用 history 子命令查看镜像的构建历史

镜像通常由多层文件构建，可以使用 history 子命令查看构建的历史，示例命令如下：

```
1 $ docker history --help
    Usage: docker history [OPTIONS] IMAGE
    Show the history of an image
    Options:
        --format string      Pretty-print images using a Go template
        -H, --human          Print sizes and dates in human readable format (default true)
        --no-trunc           Don't truncate output
        -q, --quiet          Only show image IDs

2 $ docker history centos:latest
    IMAGE           CREATED         CREATED BY                                          SIZE    COMMENT
    5d0da3dc9764    14 months ago   /bin/sh -c #(nop) CMD ["/bin/bash"]                 0B
    <missing>       14 months ago   /bin/sh -c #(nop) LABEL org.label-schema.sc…        0B
    <missing>       14 months ago   /bin/sh -c #(nop) ADD file:805cb5e15fb6e0bb0…       231MB

3 $ sudo docker history centos:latest --no-trunc
……
```

默认会截断过长的输出，可以通过 no-trunc 选项来显示完整的内容。

8.6 在 Docker 官方仓库中搜寻镜像

在实际工作中建议使用 Docker 官方的镜像。Docker 官方仓库的网址是 https://hub.docker.com/，除了通过 Web 浏览器来进行搜索之外，还可以通过 search 子命令搜索，示例命令如下：

```
$ docker search --help
  Usage:   docker search [OPTIONS] TERM
  Search the Docker Hub for images
  Options:
    -f, --filter filter     Filter output based on conditions provided
    --format string         Pretty-print search using a Go template
    --limit int             Max number of search results (default 25)
    --no-trunc              Don't truncate output
```

- -f, --filter filter：过滤输出内容。
- --format string：格式化输出内容。
- --limit int：限制输出结果的数量，默认为 25 个。
- --no-trunc：不截断输出结果。

如果希望搜索 Nginx 容器，示例命令如下：

```
1 $ docker search nginx
    NAME                DESCRIPTION                                       STARS    OFFICIAL   AUTOMATED
    nginx               Official build of Nginx.                          17691    [OK]
    linuxserver/nginx   An Nginx container, brought to you by LinuxS…     180
    bitnami/nginx       Bitnami nginx Docker Image                        142      [OK]
```

```
    ubuntu/nginx          Nginx, a high-performance reverse proxy & we...   65    ......
    ibmcom/nginx-ppc64le    Docker image for nginx-ppc64le                  0

2 docker search nginx --filter is-official=true
   AME      DESCRIPTION                STARS     OFFICIAL    AUTOMATED
   ginx     Official build of Nginx.    17691     [OK]
```

第 1 行命令的输出显示多个包含关键字 nginx 的镜像的信息，包括镜像名字、描述、星星数(也就是收藏数，表示该镜像的受欢迎程度)、是否官方创建、是否自动创建等。输出结果默认按照星级评价进行排序。

在第 2 行命令中使用了过滤条件，只显示 Nginx 官方提供的镜像。

8.7 删除和清理镜像

8.7.1 镜像的状态

在删除与清理镜像之前，需要先了解容器和镜像的状态，示例命令如下：

```
1 $ docker ps -a
    CONTAINER ID    IMAGE        COMMAND    CREATED               STATUS             PORTS    NAMES
    a1f5a5245d22    ubuntu       "bash"     27 seconds ago        Up 26 seconds      ck_bain
    2d317518f645    ubuntu:18.04  "bash"    About a minute ago    Exited (0) About
                                                                   a minute ago      tl_cdse
2 $ docker images
    REPOSITORY    TAG       IMAGE ID        CREATED           SIZE
    my-image      latest    71868614f684    4 seconds ago     77.8MB
    <none>        <none>    8a15523b6c8a    19 seconds ago    77.8MB
    ubuntu        latest    d2e4e1f51132    2 weeks ago       77.8MB
    ubuntu        18.04     c6ad7e71ba7d    2 weeks ago       63.2MB
    hello-world   latest    feb5d9fea6a5    7 months ago      13.3kB
```

(1) 正在使用的镜像。

容器会在镜像层上面增加一个读写层，它不能脱离镜像而独立存在。不管容器是否处于运行状态，都需要使用镜像。

第 2 行命令可以列出所有的容器，不论它的状态如何。

输出的第 1 行，ID 为 a1f5a5245d22 容器的状态是 Up 26 seconds，表示这个容器处于运行状态，已经运行了 26 秒。

输出的第 2 行，ID 为 2d317518f645 容器的状态是 Exited (0) About a minute ago，表示大约在 1 分钟前退出，退出代码是 0，表示是正常退出。

这台宿主机上的 ubuntu:latest 和 ubuntu:18.04 的镜像处于被使用状态，即使可以使用强制删除选项进行删除，建议也不要删除。

(2) 未使用的镜像。

没有与任何容器关联的镜像，可以安全地删除。对比第 1 行、第 2 行命令的输出，我们会发现 my-image 和 hello-world 就是未使用的镜像。删除未使用的镜像，可以释放它们所占用的存储空间。

(3) 虚悬(dangling)的镜像。

第 2 行命令的输出中显示：ID 是 8a15523b6c8a 的镜像既没有 REPOSITORY，也没有 TAG。这类镜像被称为虚悬镜像，属于非正常状态的镜像。导致出现虚悬镜像的原因有多种，在后续的章节会有详细介绍。

8.7.2 删除镜像

可以使用 rmi 或 image rm 子命令删除镜像，这两个命令是等效的，都是从宿主机中删除一个或多个镜像。

```
docker rmi [OPTIONS] IMAGE [IMAGE...]
docker image rm [OPTIONS] IMAGE [IMAGE...]
```

它们的选项也相同：

- -f,--force：强制删除镜像，不论是否有容器依赖。
- --no-prune：不要清理未带标签的父镜像。

在删除时，会先进行 UNTAG(取消标记)操作。如果这个镜像有多个 TAG，则仅会删除指定的 TAG，而不会删除宿主机中的文件。如果这个镜像只有一个 TAG，在删除 TAG 的同时，还会删除宿主机中的文件。这类似于删除文件系统的硬连接。

在删除时，既可以指定名称+TAG，也可以指定镜像的 ID，示例命令如下：

```
1 $ docker images
  REPOSITORY      TAG      IMAGE ID       CREATED         SIZE
  test1           latest   feb5d9fea6a5   7 months ago    13.3kB
  test2           latest   feb5d9fea6a5   7 months ago    13.3kB
  hello-world     latest   feb5d9fea6a5   7 months ago    13.3kB

2 $ docker rmi feb5d9fea6a5
  Error response from daemon: conflict: unable to delete feb5d9fea6a5 (must be forced) - image is referenced in
  multiple repositories

3 $ docker rmi test2
  Untagged: test2:latest

4 $ docker rmi -f feb5d9fea6a5
  Untagged: hello-world:latest
  Untagged: hello-world@sha256:80f31da1ac7b312ba29d65080fddf797dd76acfb870e677f390d5acba9741b17
  Untagged: test1:latest
  Deleted: sha256:feb5d9fea6a5e9606aa995e879d862b825965ba48de054caab5ef356dc6b3412
  Deleted: sha256:e07ee1baac5fae6a26f30cabfe54a36d3402f96afda318fe0a96cec4ca393359

5 $ docker images
  REPOSITORY      TAG      IMAGE ID      CREATED      SIZE
```

第 1 行命令的输出显示：当前宿主机有 3 个镜像。由于它们的 ID 相同，所以这 3 个镜像是同一个镜像，只不过拥有不同的 REPOSITORY+TAG 而已。

第 2 行命令使用 ID 进行删除，会收到错误提示，不允许删除具有多个引用的镜像。

第 3 行命令使用标签进行删除，这仅仅是删除一个标签，所以输出显示的操作是 Untagged，说明还有其他镜像在引用它。

第 4 行命令使用-f 选项进行强制删除，命令的输出显示：先进行多次 UNTAG 操作，当所有的 TAG 被删除后，才会删除/var/lib/docker/overlay2/目录的相应文件。

8.7.3 清理镜像

使用 Docker 一段时间后，系统中可能会遗留一些临时的、无用的镜像文件，例如，虚悬的镜像、没有被使用的镜像。可以使用 image prune 子命令进行清理以释放占用的存储空间，它有 3 个选项：

- -a,--all：删除所有虚悬镜像和未使用的镜像，默认仅删除所有虚悬镜像。
- --filter, filter：只清理符合指定条件的镜像，例如通过指定时间来删除多长时间没有被使用过的镜像。
- -f,--force：强制删除镜像，不进行提示确认。

示例命令如下：

```
1 $ docker ps -a
    CONTAINER ID   IMAGE   COMMAND   CREATED   STATUS   PORTS   NAMES

2 $ docker image ls
    REPOSITORY   TAG      IMAGE ID       CREATED         SIZE
    <none>       <none>   bfe296a52501   7 days ago      5.54MB
    centos       latest   5d0da3dc9764   14 months ago   231MB

3 $ docker image prune
    WARNING! This will remove all dangling images.
    Are you sure you want to continue? [y/N] y
    Deleted Images:
    deleted: sha256:bfe296a525011f7eb76075d688c681ca4feaad5afe3b142b36e30f1a171dc99a
    deleted: sha256:e5e13b0c77cbb769548077189c3da2f0a764ceca06af49d8d558e759f5c232bd

    Total reclaimed space: 5.544MB
4 $ docker image ls
    REPOSITORY   TAG      IMAGE ID       CREATED         SIZE
    centos       latest   5d0da3dc9764   14 months ago   231MB
```

image prune 子命令默认只清理所有虚悬的镜像。第 4 行命令的输出显示：系统还保留着 centos:latest 镜像。

```
5 $ docker image prune -a
    WARNING! This will remove all images without at least one container associated to them.
    Are you sure you want to continue? [y/N] y
    Deleted Images:
    untagged: centos:latest
    untagged: centos@sha256:a27fd8080b517143cbbbab9dfb7c8571c40d67d534bbdee55bd6c473f432b177
    deleted: sha256:5d0da3dc976460b72c77d94c8a1ad043720b0416bfc16c52c45d4847e53fadb6
    deleted: sha256:74ddd0ec08fa43d09f32636ba91a0a3053b02cb4627c35051aff89f853606b59
```

```
    Total reclaimed space: 231.3MB
6 $ docker image ls
    REPOSITORY    TAG        IMAGE ID    CREATED    SIZE

7 $ sudo ls -l /var/lib/docker/overlay2
    total 0
    brw-------. 1 root root 253, 0 Nov 20    07:47   backingFsBlockDev
    drwx------. 2 root root        6 Nov 20    08:30   l
```

第 5 行命令使用-a 选项，将删除所有未使用的镜像。

第 7 行命令的输出显示：在/var/lib/docker/overlay2 目录中，已经没有镜像文件了。

8.8 创建新镜像

因为官方镜像只提供基础的功能，所以在生产中通常在它们的基础上创建新的镜像。主要有 3 种方法：

- 基于已有容器创建新镜像，需要使用 docker commit 命令。
- 使用 Dockerfile 创建新镜像，需要使用 docker build 命令。
- 通过导入本地模板创建新镜像，需要使用 docker import 命令。

8.8.1 基于已有容器创建新镜像

容器在只读镜像上添加了一个读写层，用户的添加、修改、删除等操作都被保存在这一块层中。docker commit 命令会将容器的读写层固化下来，永久保存为新镜像，如图 8-10 所示。

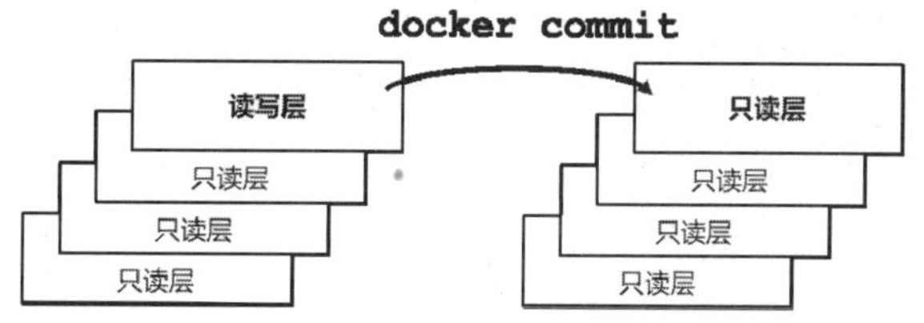

图 8-10 docker commit 的工作原理

在执行 commit 子命令时，如果容器处于运行状态，Docker 为了保证数据的一致性，默认情况下会将容器先短暂地“静默”一下，以便将缓存中的数据写入存储，然后再创建新的镜像。

基于已有容器创建新镜像的过程包括以下 3 个步骤：

(1) 运行容器。

(2) 在容器中进行文件修改、软件安装等操作。

(3) 使用 docker commit 命令创建新的镜像。

下面以 Alpine Linux 容器为基础创建一个新的镜像，示例命令如下：

```
1 $ docker pull alpine

2 $ docker images
```

```
    REPOSITORY  TAG       IMAGE ID         CREATED        SIZE
    alpine      latest    bfe296a52501     8 days ago     5.54MB

3 $ sudo ls -l /var/lib/docker/overlay2/
    total 0
    drwx--x---. 3 root root 30 Nov 20 15:03
    91ffe10e9a665d9da63425951d297723ff528f22553db46b185f702220a8907c
    brw-------. 1 root root 253, 0 Nov 20 14:16 backingFsBlockDev
    drwx------. 2 root root 40 Nov 20 15:03 l
```

Alpine Linux 是一个社区开发的面向安全应用的轻量级 Linux 发行版，体积小，特别适用于 Docker 镜像、路由器、防火墙、VPN、VoIP 盒子等业务。

```
4 $ docker run -it alpine sh
    / # cat /etc/os-release
    NAME="Alpine Linux"
    ID=alpine
    VERSION_ID=3.16.3
    PRETTY_NAME="Alpine Linux v3.16"
    HOME_URL="https://alpinelinux.org/"
    BUG_REPORT_URL="https://gitlab.alpinelinux.org/alpine/aports/-/issues"
    / # pwd
    /
    / # date > test.txt
    / # exit

5 $ docker ps -a
    CONTAINER ID  IMAGE   COMMAND   CREATED              STATUS                     PORTS   NAMES
    ceb3c847b4c2  alpine  "sh"      About a minute ago   Exited (0) 14 seconds ago
                                                                                            modest_neumann
```

执行第 4 行命令运行容器并进入到 Shell 环境，查看发行版本的信息，在根目录下创建测试文件 test.txt，然后使用 exit 命令退出容器。

第 5 行命令的输出显示容器状态是 Exited (0)，表示为正常退出。

```
6 $ sudo ls -l /var/lib/docker/overlay2/
    total 0
    drwx--x---. 4 root root 55 Nov 20 15:15
              20d6cab6adc73e33d4e15a2c42c6f5d08288192c574d55d37f7a31779163a520
    drwx--x---. 4 root root 72 Nov 20 15:14
              20d6cab6adc73e33d4e15a2c42c6f5d08288192c574d55d37f7a31779163a520-init
    drwx--x---. 3 root root 47 Nov 20 15:14
              91ffe10e9a665d9da63425951d297723ff528f22553db46b185f702220a8907c
    brw-------. 1 root root 253, 0 Nov 20 14:16 backingFsBlockDev
    drwx------. 2 root root 108 Nov 20 15:14 l

7 $ sudo tree
    /var/lib/docker/overlay2/20d6cab6adc73e33d4e15a2c42c6f5d08288192c574d55d37f7a31779163a520/
    /var/lib/docker/overlay2/20d6cab6adc73e33d4e15a2c42c6f5d08288192c574d55d37f7a31779163a520/
     ├── diff
     │   ├── root
```

```
|   └──── test.txt
├──── link
├──── lower
└──── work
    └──── work
```

第 6 行命令的输出显示：在/var/lib/docker/overlay2/目录中新增了两个子目录，它们共同组成了镜像的读写层，其中带 init 的目录对容器是只读的，没有 init 的容器目录是容器的读写目录。从第 7 行的命令输出可以看出：写入容器的文件都保存在这个目录中。

```
8 $ docker commit -m "Add new files" -a "Tom" ceb3c847b4c2 mytest:0.1
    sha256:fef5f865b17747383437e46a33ab7e947cdbb63c2442b75d584374a85d4db75b
```

commit 子命令的选项包括：

- -a, --author=""：作者信息。
- -c, --change=：提交时执行的 Dockerfile 格式的指令。
- -m, --message=""：注释信息。
- -p, --pause[=true]：提交时暂停容器运行，从而保证数据的一致性，默认为 true。

```
9 $ docker images
    REPOSITORY    TAG       IMAGE ID        CREATED          SIZE
    mytest        0.1       fef5f865b177    9 seconds ago    5.54MB
    alpine        latest    bfe296a52501    8 days ago       5.54MB

10 $ sudo ls -l -t /var/lib/docker/overlay2
    total 0
    drwx------. 2 root root 142 Nov 20 17:05 l
    drwx--x---. 4 root root 72 Nov 20 15:25
    6ea65b97b434074bf69fd2ba233e00235a89e91a28caef96dd5f1cb98d38a20d
    drwx--x---. 4 root root 55 Nov 20 15:22
    20d6cab6adc73e33d4e15a2c42c6f5d08288192c574d55d37f7a31779163a520
    drwx--x---. 4 root root 72 Nov 20 15:14
    20d6cab6adc73e33d4e15a2c42c6f5d08288192c574d55d37f7a31779163a520-init
    drwx--x---. 3 root root 47 Nov 20 15:14
    91ffe10e9a665d9da63425951d297723ff528f22553db46b185f702220a8907c
    brw-------. 1 root root 253, 0 Nov 20 14:16 backingFsBlockDev
```

第 9 行命令的输出显示了新镜像的信息。通过第 10 行命令可以看出/var/lib/docker/overlay2 目录中增加了新镜像的目录 6ea65b97b434074bf69fd2ba233e00235a89e91a28caef96dd5f1cb98d38a20d。

```
11 $ docker run -it mytest:0.1 sh
    / # cat test.txt
    Sun Nov 20 07:15:09 UTC 2022
    / # exit
```

通过第 11 行命令基于自定义镜像创建新的容器，进入容器之后，可以在根目录下看到文件 test.txt 的内容，这说明自定义镜像创建成功。

8.8.2 使用 Dockerfile 创建新镜像

Docker 的 commit 子命令是基于已有容器来创建镜像，那么最初的镜像是怎么来的呢？由创建实现，有一种做法是使用 build 子命令读取 Dockerfile 中的指令来构建新的镜像，如图 8-11 所示。

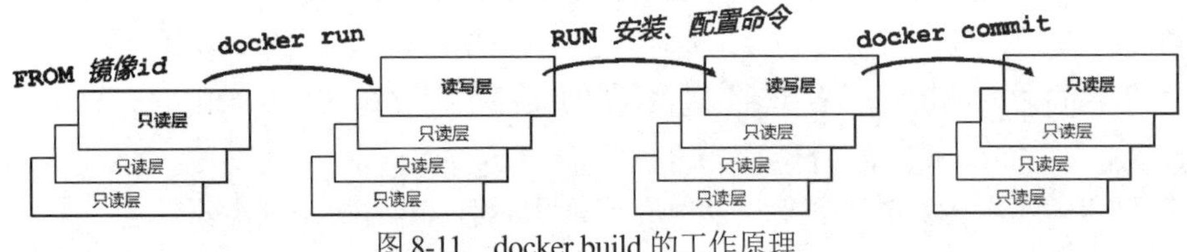

图 8-11　docker build 的工作原理

(1) Dockerfile 是一个包含镜像构建指令的文本文件。通过它，既可以开天辟地、从无到有地创建镜像，也可以基于某个父镜像创建新的子新镜像。Dockerfile 的 FROM 指令用于指定基础镜像，它必须是第一个指令。scratch 是特殊的基础镜像，是一个空白的镜像，用于创建一个全新的镜像。

(2) build 子命令首先会调用 run 子命令，根据基础镜像创建一个临时的容器。

(3) 根据 Dockerfile 中的 RUN 指令在容器中进行安装、配置操作，这些操作会写入到容器的读写层。

(4) build 子命令再调用 commit 子命令，根据临时容器创建新的镜像，创建完成后删除临时容器。

使用 8.8.1 节中的方法对容器的修改是手工操作，不便于对操作过程进行跟踪记录。根据 Dockerfile 中的指令进行操作，对容器的修改是自动化完成的，操作的过程可追溯，也方便排错。

提示

官方镜像仓库 https://hub.docker.com/提供构建每个镜像文件的 Dockerfile，这是很好的参考资料。

下面，通过 3 个示例简单介绍 Dockerfile 的指令。

(1) Hello-world 镜像。

```
FROM scratch
# FROM 指令用于指定基础镜像，它必须为第一个指令。使用 scratch 镜像，可以从零开始构建。
COPY hello /
#复制本地文件 hello 到镜像的根目录。COPY 指令类似 ADD 指令，但不会自动解压文件，也不能访问网
 络资源。
CMD ["/hello"]
#当容器启动时，运行根目录下的 hello 文件。
```

(2) Ubuntu 镜像。

```
FROM scratch
ADD ubuntu-kinetic-oci-amd64-root.tar.gz /
#复制本地 TAR 格式的 Ubuntu 发行版本的文件到镜像的根目录，ADD 指令会自动解压缩这个文件。
```

```
#ADD 指令还可以像 wget 命令一样访问网络资源。
CMD ["bash"]
#当容器启动时，运行 bash 构建一个 Shell 环境。
```

(3) HTTPD 镜像复杂很多，需要进行安装软件包、修改配置等多种操作。

```
FROM debian:bullseye-slim
# 基于 Debian 镜像。
# add our user and group first to make sure their IDs get assigned consistently, regardless of whatever dependencies
  get added
#RUN groupadd -r www-data && useradd -r --create-home -g www-data www-data

# #号是注释说明的起始符号。

# ENV 指令用于设置环境变量。
ENV HTTPD_PREFIX /usr/local/apache2
ENV PATH $HTTPD_PREFIX/bin:$PATH

# RUN 指令是构建镜像时执行的命令，例如，创建目录、修改权限等，其中可以使用环境变量。
RUN mkdir -p "$HTTPD_PREFIX" \
    && chown www-data:www-data "$HTTPD_PREFIX"
# WORKDIR 指令可以切换工作目录，类似于 cd 命令。
WORKDIR $HTTPD_PREFIX

# install httpd runtime dependencies
# https://httpd.apache.org/docs/2.4/install.html#requirements
RUN set -eux; \
    apt-get update; \
    apt-get install -y --no-install-recommends \
        libaprutil1-ldap \
# https://github.com/docker-library/httpd/issues/209
        libldap-common \
    ; \
rm -rf /var/lib/apt/lists/*
......
```

下面介绍如何“复刻”hello-world 镜像，示例命令如下：

```
1 $ mkdir test && cd test

2 $ wget https://raw.githubusercontent.com/docker-library/hello-world/master/i386/hello-world/hello

3 $ file hello
   hello: ELF 32-bit LSB executable, Intel 80386, version 1 (SYSV), statically linked, stripped
```

build 子命令将上传构建目录中的所有文件，所以建议在全新的目录中构建镜像。

在 https://hub.docker.com/上搜索 hello-world，根据提示找到编译好的 hello 二进制文件，下载到当前目录中。

第 3 行命令的输出显示：hello 是静态链接型的可执行文件，这种格式不依赖其他库文件即可执行。如果希望下载 hello.c 源代码文件进行编译，在编译时需要指定静态链接选项。

```
4 $ chmod +x hello
```

```
5 $ ./hello
    Hello from Docker!
    This message shows that your installation appears to be working correctly.
    ......
```

第 4 行命令给文件添加可执行权限，第 5 行命令检查执行的情况。

```
6 $ vi Dockerfile
    FROM scratch
    COPY hello /
    CMD ["/hello"]

7 $ docker build -t myhello:1.0 .
    Sending build context to Docker daemon    15.87kB
    Step 1/3 : FROM scratch
     --->
    Step 2/3 : COPY hello /
     ---> 4f03800fe966
    Step 3/3 : CMD ["/hello"]
     ---> Running in 1775b52a6de0
    Removing intermediate container 1775b52a6de0
     ---> 9e8e4cdf0ba2
    Successfully built 9e8e4cdf0ba2
    Successfully tagged myhello:1.0
```

第 7 行命令的输出显示：本次构建分为 3 个步骤，第 1 步是启动临时的 scratch 容器，第 2 步是向其中复制文件 hello，第 3 步是修改容器的 CMD 属性。在 commit 新镜像(ID 为 9e8e4cdf0ba2)之后，再删除中间的临时容器(ID 为 1775b52a6de0)，最后给新镜像添加标签 myhello:1.0。

```
8 $ docker images
    REPOSITORY    TAG          IMAGE ID        CREATED          SIZE
    myhello       1.0          9e8e4cdf0ba2    19 seconds ago   12.9kB

9 $ docker image history myhello:1.0
    IMAGE           CREATED          CREATED BY                                        SIZE COMMENT
    9e8e4cdf0ba2    28 seconds ago   /bin/sh -c #(nop) CMD ["/hello"]                  0B
    4f03800fe966    28 seconds ago   /bin/sh -c #(nop) COPY file:0521034fa7e53a3e...   12.9kB
```

第 9 行命令显示了新镜像的构建历史信息，它与第 7 行命令的输出一一对应。

```
10 $ sudo docker run myhello:1.0
    Hello from Docker!
    This message shows that your installation appears to be working correctly.
    ......
```

8.8.3 导入本地模板来创建新镜像

如果需要把一个运行良好的容器移植到多台宿主机上，在没有公共或私有的镜像仓库的情况下，可以使用导出导入的方法来创建新镜像。

首先使用 export 子命令在源宿主机上将容器导出为 TAR 格式的归档文件(也可以认为是模

板文件)，然后在目标宿主机上使用 import 子命令进行导入，从而创建新镜像，如图 8-12 所示。

图 8-12　docker export/import 的工作原理

export 子命令的语法格式如下：

```
docker export [OPTIONS] CONTAINER
```

用法：将容器的文件系统导出为 TAR 格式的存档文件。

选项：

- -o, --output string：指定导出到的文件，默认是导出到标准输出(STDOUT)。

import 子命令的语法格式如下：

```
docker import [OPTIONS] file|URL|- [REPOSITORY[:TAG]]
```

用法：导入 TAR 格式的存档文件内容以创建新镜像。

选项：

- -c, --change list：在镜像中添加 Dockerfile 格式的指令。
- -m, --message string：为镜像设置注释信息。
- --platform string：如果服务器支持多平台，则设置平台的信息。

下面做导出导入镜像的实验，示例命令如下：

```
1 $ mkdir test && cd test

2 $ docker run -it alpine sh
    / # date > new.txt
    / # exit
```

第 2 行命令运行一个 Alpine Linux 容器，并在其中创建一个测试文件。

```
3 $ docker images -a
    REPOSITORY    TAG       IMAGE ID        CREATED       SIZE
    alpine        latest    49176f190c7e    2 days ago    7.05MB

4 $ docker ps -a
    CONTAINER ID  IMAGE    COMMAND   CREATED            STATUS                      PORTS
    3c35d9023839  alpine   "sh"      31 seconds ago     Exited (0) 19 seconds ago   keen_spence
                                                                                    NAMES

5 $ docker export 3c35d9023839 -o export.tar

6 $ ls -lh
    total 7.1M
    -rw-------. 1 tom tom 7.1M Nov 25 11:11 export.tar
```

第 5 行命令导出容器，既可以指定容器的 ID，也可以使用容器的名称。导出的文件是已经合并的容器的文件系统。下面将它解压缩到一个新目录中，查看其中的文件。

```
7 $ mkdir exptest

8 $ tar -xf export.tar -C exptest/

9 $ tree -L 1 exptest/
    exptest/
    ├── bin
    ├── dev
    ├── etc
    ├── home
    ├── lib
    ├── media
    ├── mnt
    ├── new.txt
    ……
```

通过导入创建新的镜像，示例命令如下：

```
10 $ sudo docker import -m "Import test" export.tar myalpine:1.0
    sha256:7391ae07119f0d48ab114b0fa00e96138e993c2203b876193308cada08ae9809

11 $ docker image ls
    REPOSITORY  TAG      IMAGE ID      CREATED         SIZE
    myalpine    1.0      7391ae07119f  7 seconds ago   7.05MB
    alpine      latest   49176f190c7e  2 days ago      7.05MB

12 $ docker history myalpine:1.0
    IMAGE         CREATED          CREATED BY   SIZE     COMMENT
    7391ae07119f  13 seconds ago                7.05MB   Import test

13 $ docker run myalpine:1.0 cat /new.txt
    Fri Nov 25 03:09:45 UTC 2022
```

第 10 行命令在进行导入时，还设置了可选的注释信息。

第 12 行命令的输出显示：新镜像只有一层。

第 13 行命令运行新镜像时，直接使用 cat 命令查看测试文件是否在。

8.9 保存与加载镜像

虽然保存(save)/加载(load)镜像与容器导出(export)/导入(import)有些相似，但是它们的作用机理是不一样的：

- export/import 子命令操作的对象是容器。它会把容器和镜像的各层合并在一起，然后打包形成一个 TAR 格式的文件。

- save/load 子命令操作的对象是镜像。虽然也是打包形成一个 TAR 格式的文件，但是保留了镜像各个层的文件及其所有元数据。

保存与加载镜像主要的应用场景是镜像的备份与恢复，可以减少下载镜像的时间。其工作原理如图 8-13 所示。

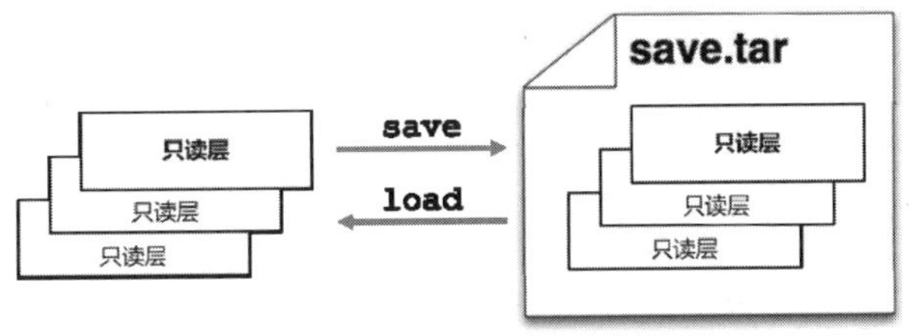

图 8-13 save/load 镜像的工作原理

save 子命令的语法格式如下：

```
docker image save [OPTIONS] IMAGE [IMAGE...]
  docker save [OPTIONS] IMAGE [IMAGE...]
```

用法：将一个或多个镜像保存到 TAG 格式归档文件(默认情况下传输到 STDOUT)。

选项：

- -o, --output string：指定导出到的文件。

load 子命令的语法格式如下：

```
docker load [OPTIONS]
docker image load [OPTIONS]
```

用法：从 TAR 格式的存档文件或 STDIN 加载镜像。

选项：

- -i, --input string：指定读取的文件。
- -q, --quiet：不显示加载时的输出。

假设有这样一个场景：一台宿主机上有多个镜像，为了避免误操作，需要进行备份，操作命令如下：

```
1 $ mkdir test && cd test

2 $ docker images
    REPOSITORY     TAG       IMAGE ID        CREATED          SIZE
    alpine         latest    49176f190c7e    3 days ago       7.05MB
    busybox        latest    9d5226e6ce3f    8 days ago       1.24MB
    mysql          latest    3842e9cdffd2    9 days ago       538MB
    httpd          latest    8653efc8c72d    10 days ago      145MB
    nginx          latest    88736fe82739    10 days ago      142MB
    ubuntu         latest    a8780b506fa4    3 weeks ago      77.8MB
    hello-world    latest    feb5d9fea6a5    14 months ago    13.3kB
    centos         latest    5d0da3dc9764    14 months ago    231MB

3 $ docker image save -o save.tar alpine busybox mysql httpd nginx ubuntu hello-world centos
```

save 子命令可以保存多个镜像，通过-o 选项指定要创建的文件名。

考察归档文件中的内容，示例命令如下：

```
4 $ ls -lh
    total 1.1G
    -rw-------. 1 tom tom 1.1G Nov 26 09:03 save.tar

5 $ mkdir untar

6 $ tar -xf save.tar -C untar/

7 $ ls untar/
    079bc5e75545bf45253ab44ce73fbd51d96fa52ee799031e60b65a82e89df662
    1f0acadf10311f425cc1bc876cb6ff5fb9e20064181b2168c1e1ad2a105b6055
    22518b5865066299ddaadda6c5bdc413668e1eb33ef842962e2b7ab93a0cb781
    268cade3441253bda04666348964e6217ba79c064b8ef9ef1eca3e650ee5454c
    3842e9cdffd239649671beaec03b363b52f7b050fbb4a8869c052438408d6d2e.json
    ……
    feb5d9fea6a5e9606aa995e879d862b825965ba48de054caab5ef356dc6b3412.json
    manifest.json
    repositories
```

第 7 行命令的输出显示：在归档文件中不仅有每个镜像的各层文件，还有多个 JSON 格式的描述文件，它们一起用于将来的加载操作。

```
8 $ docker image prune -a -f
```

假设某个用户误操作，使用第 8 行命令强制清除了所有的镜像。这时，就可以通过 load 子命令进行恢复，示例命令如下：

```
9 $ docker image load -i save.tar
    74ddd0ec08fa: Loading layer    238.6MB/238.6MB
    Loaded image: centos:latest
    ded7a220bb05: Loading layer    7.338MB/7.338MB
    Loaded image: alpine:latest
    40cf597a9181: Loading layer    1.459MB/1.459MB
    Loaded image: busybox:latest
    4d0c6342b0f5: Loading layer    109.2MB/109.2MB
    437bfe664fb2: Loading layer    11.26kB/11.26kB
    b86bd80e1568: Loading layer    2.293MB/2.293MB
    1e4614d1f65d: Loading layer    13.92MB/13.92MB
    ed5c1da221b6: Loading layer    7.168kB/7.168kB
    ef49e07a76fb: Loading layer    3.072kB/3.072kB
    6b7e563dc9ef: Loading layer    178.4MB/178.4MB
    d785c41e315b: Loading layer    3.072kB/3.072kB
    8848f79a3581: Loading layer    246.5MB/246.5MB
    615e49ac424e: Loading layer    17.41kB/17.41kB

10 $ docker images
    REPOSITORY    TAG      IMAGE ID        CREATED       SIZE
    alpine        latest   49176f190c7e    3 days ago    7.05MB
    busybox       latest   9d5226e6ce3f    8 days ago    1.24MB
```

```
mysql          latest    3842e9cdffd2    9 days ago      538MB
httpd          latest    8653efc8c72d    10 days ago     145MB
nginx          latest    88736fe82739    10 days ago     142MB
ubuntu         latest    a8780b506fa4    3 weeks ago     77.8MB
hello-world    latest    feb5d9fea6a5    14 months ago   13.3kB
centos         latest    5d0da3dc9764    14 months ago   231MB
```

第 9 行命令使用-i --input string 选项，指定要读取的文件名称。第 10 行命令输出显示：所有的镜像都恢复了。

8.10 集中管理镜像

在生产环境中，会大量使用自定义镜像，这就需要集中化管理自定义镜像，主要有两种管理自定义镜像的方法：

- 通过 Docker 等公司的公共仓库进行管理。
- 通过私有镜像仓库进行管理。

8.10.1 上传镜像到公共仓库

如果镜像是可以公开的，那么就使用 Docker 等公司的公共仓库。Docker 命令的上传仓库默认是 Docker 的官方仓库。在 Docker Hub 上注册账户之后，就可以上传镜像，如图 8-14 所示。

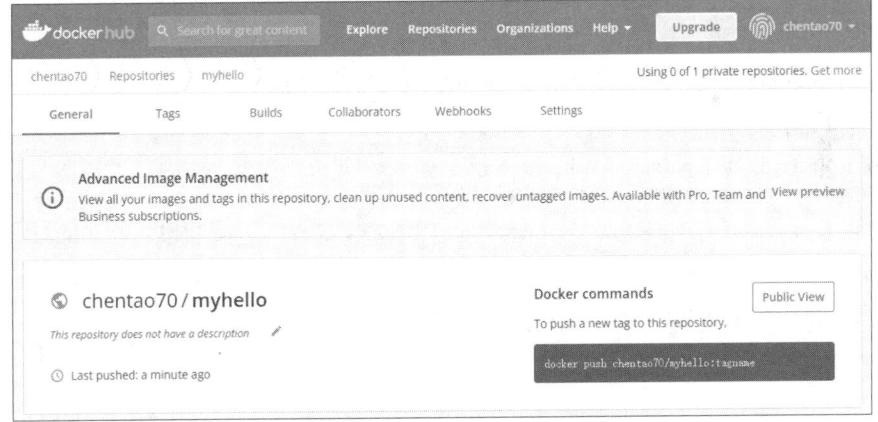

图 8-14　Docker Hub 上的个人镜像

可以使用 push 子命令上传镜像，它的语法格式如下：

```
docker image push [OPTIONS] NAME[:TAG]
docker push [OPTIONS] NAME[:TAG]
```

用法：将镜像或存储库(repository)推送到注册表(registry)。

选项：

- -a, --all-tags：将所有标签的镜像推送到存储库中。
- --disable-content-trust：禁用内容信任，跳过镜像签名(默认为 true)。

- -q, --quiet：抑制详细输出。

下面做一个上传镜像到公共仓库的实验。在上传镜像之前，需要使用 login 子命令配置身份凭证，示例命令如下：

```
1 $ docker login
   Login with your Docker ID to push and pull images from Docker Hub. If you don't have a Docker ID, head
   over to https://hub.docker.com to create one.
   Username: chentao70
   Password: ********  输入密码
   WARNING! Your password will be stored unencrypted in /home/tom/.docker/config.json.
   Configure a credential helper to remove this warning. See
   https://docs.docker.com/engine/reference/commandline/login/#credentials-store

   Login Succeeded
```

准备自定义镜像(通过给现有镜像添加新标签的方式来创建新镜像)，示例命令如下：

```
2 $ docker images
   REPOSITORY        TAG       IMAGE ID        CREATED          SIZE
   hello-world       latest    feb5d9fea6a5    14 months ago    13.3kB

3 $ docker tag hello-world:latest chentao70/myhello:1.0

4 $ docker images
   REPOSITORY          TAG       IMAGE ID        CREATED          SIZE
   chentao70/myhello   1.0       feb5d9fea6a5    14 months ago    13.3kB
   hello-world         latest    feb5d9fea6a5    14 months ago    13.3kB
```

Docker Hub 为了区分不同用户的镜像，要求自定义镜像名称必须要包含用户名的信息，格式为：[username]/xxx:tag。所以第 3 行命令添加的标签为 chentao70/myhello:1.0。

```
5 $ docker push chentao70/myhello:1.0
   The push refers to repository [docker.io/chentao70/myhello]
   e07ee1baac5f: Mounted from library/hello-world
   1.0: digest: sha256:f54a58bc1aac5ea1a25d796ae155dc228b3f0e11d046ae276b39c4bf2f13d8c4 size: 525
```

上传成功之后，可以立即在 Docker Hub 的网站上看到新镜像的信息，如图 8-15 所示。

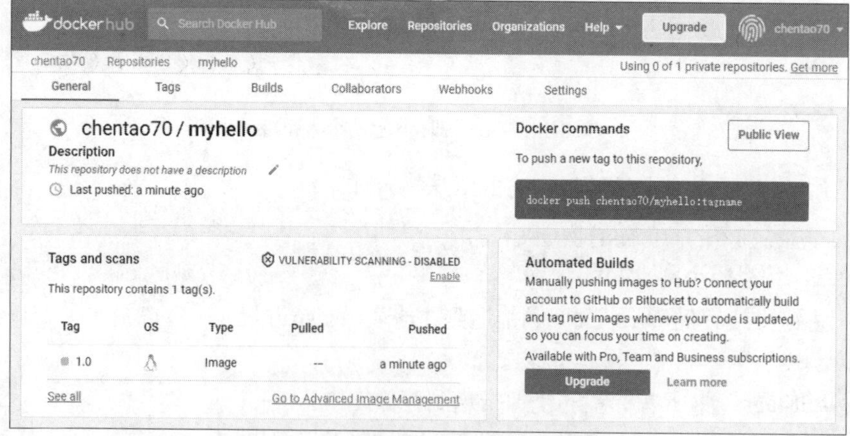

图 8-15　新上传的个人镜像的详细信息

还可以通过 search 子命令搜索自定义的镜像信息(通常需要稍等一会才能搜索到)，示例命令如下：

```
6 $ docker search chentao70
    NAME                    DESCRIPTION   STARS     OFFICIAL   AUTOMATED
    chentao70/myhello                     0
```

清除本地实验环境中的现有镜像，然后测试新上传的自定义镜像是否可以使用，示例命令如下：

```
7 $ docker image prune -a -f

8 $ docker run chentao70/myhello:1.0
    Unable to find image 'chentao70/myhello:1.0' locally
    1.0: Pulling from chentao70/myhello
    2db29710123e: Pull complete
    Digest: sha256:f54a58bc1aac5ea1a25d796ae155dc228b3f0e11d046ae276b39c4bf2f13d8c4
    Status: Downloaded newer image for chentao70/myhello:1.0

    Hello from Docker!
    This message shows that your installation appears to be working correctly.
    ……

9 $ docker images
    REPOSITORY          TAG      IMAGE ID       CREATED         SIZE
    chentao70/myhello   1.0      feb5d9fea6a5   14 months ago   13.3kB
```

8.10.2 上传镜像到私有仓库

私有镜像仓库的应用场景有很多，其中一个典型的场景是企业内网用户使用 Docker，如图 8-16 所示。

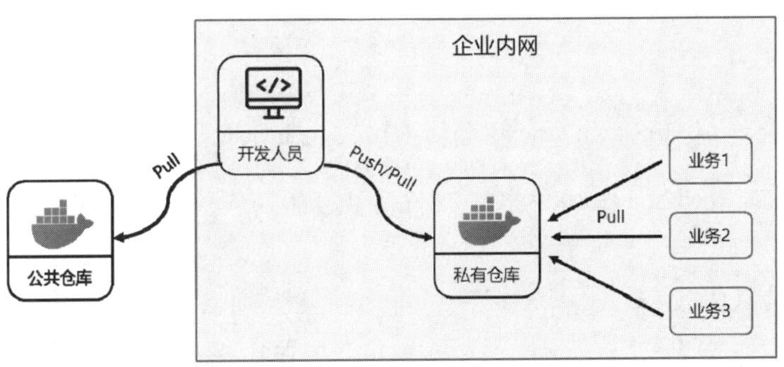

图 8-16　企业内网用户使用 Docker 的场景

开发人员从公共仓库下载镜像进行开发工作，然后把业务数据、配置等信息保存在自定义镜像中。测试通过后，希望其他业务部门同时来使用。为了安全，需要将自定义镜像保存在企业内网中的私有镜像仓库，而不是公共的镜像仓库。这种架构还有一个优点：内网用户下载镜像的速度会比较快。

Docker 已经开源了私有镜像仓库的核心功能。最简单的使用方法是直接使用 Docker 官方提供的私有镜像仓库的镜像，它的功能可以满足大多数企业的需求。

默认情况下，此镜像会使用 HTTP 而不是 HTTPS 协议，但是 Docker 的 pull 子命令默认使用 HTTPS 协议下载镜像，在下面的实验中，我们需要修改 Docker 守护程序配置文件，将私有镜像仓库地址添加到“不安全注册表列表”中，示例命令如下：

```
1 $ sudo vi /etc/docker/daemon.json
  # 创建新文件，添加如下内容，私有仓库的 IP 地址为 192.168.1.231
  {
          "insecure-registries":[
                  "192.168.1.231:5000"
          ]
  }

2 $ sudo systemctl restart docker
```

提示

仅在实验、测试或严格控制的环境中使用此方法，强烈建议使用由可信 CA 颁发的 TLS 证书保护注册表。

运行容器，示例命令如下：

```
3 $ docker run -d -p 5000:5000 -v /labregistry:/var/lib/registry registry
```

运行时使用了 3 个选项：

- -d：在后台启动容器，只在屏幕上显示容器的 ID。
- -p：将容器的 5000 端口映射到宿主机的 5000 端口。5000 是 registry 的默认服务端口。
- -v：将容器/var/lib/registry 目录映射到宿主机的/labregistry 目录，用于存放镜像数据。这是持久存储，即使关闭或删除容器，仍然可以访问这些文件。

考察容器运行的状态，示例命令如下：

```
4 $ docker ps -a
    CONTAINER ID   IMAGE    COMMAND                  CREATED        STATUS
    62aa64f84bc6   registry "/entrypoint.sh /etc…"   5 minutes ago  Up 1 minutes
                                    PORTS                 NAMES
    0.0.0.0:5000->5000/tcp, :::5000->5000/tcp    friendly_borg

5 $ sudo ss -tlnp |grep 5000
    [sudo] password for tom:
    LISTEN 0    4096    0.0.0.0:5000    0.0.0.0:*  users:(("docker-proxy",pid=3578,fd=4))
    LISTEN 0    4096    [::]:5000       [::]:*     users:(("docker-proxy",pid=3586,fd=4))

6 $ ls -l /labregistry/
    total 0
```

创建测试用的自定义镜像，示例命令如下：

```
7 $ docker run -it alpine sh
    / # date > new.txt
    / # exit
```

```
8 $ docker ps -a
  #获得 alpine 容器 ID 用于创建新镜像，例如 01020d0ebdd9。

9 $ docker commit 01020d0ebdd9 labapp1

10 $ docker images
     REPOSITORY      TAG        IMAGE ID        CREATED          SIZE
     labapp1         latest     e1a9b9a51001    5 seconds ago    7.05MB
     alpine          latest     49176f190c7e    3 days ago       7.05MB
     registry        latest     81c944c2288b    13 days ago      24.1MB
```

配置宿主机防火墙，允许 TCP 5000 端口数据包的入站，示例命令如下：

```
11 $ sudo firewall-cmd --add-port=5000/tcp --permanent

12 $ sudo firewall-cmd --reload

13 $ sudo firewall-cmd --list-all
        public (active)
        target: default
        icmp-block-inversion: no
        interfaces: ens160
        sources:
        services: cockpit dhcpv6-client ssh
        ports: 5000/tcp
        protocols:
        forward: yes
        masquerade: no
        forward-ports:
        source-ports:
        icmp-blocks:
        rich rules:
```

镜像名称包括 repository 和 tag 两部分，而 repository 的完整格式为：[registry-host]:[port]/[username]/xxx。因为只有 Docker 官方的镜像可以省略[registry-host]:[port]，所以根据需要私有仓库的 IP 地址(主机名或域名)给镜像添加标签，示例命令如下：

```
14 $ docker images
     REPOSITORY      TAG        IMAGE ID        CREATED          SIZE
     labapp1         latest     e1a9b9a51001    6 minutes ago    7.05MB
     alpine          latest     49176f190c7e    3 days ago       7.05MB
     registry        latest     81c944c2288b    13 days ago      24.1MB

15 $ docker tag labapp1:latest 192.168.1.231:5000/oa/labapp1:latest

16 $ docker images
     REPOSITORY                      TAG        IMAGE ID        CREATED          SIZE
     192.168.1.231:5000/oa/labapp1   latest     e1a9b9a51001    6 minutes ago    7.05MB
     labapp1                         latest     e1a9b9a51001    6 minutes ago    7.05MB
     alpine                          latest     49176f190c7e    3 days ago       7.05MB
```

```
registry                    latest       81c944c2288b   13 days ago      24.1MB
```

使用 push 子命令将自定义镜像上传到私有软件仓库，示例命令如下：

```
17 $ docker push 192.168.1.231:5000/oa/labapp1
    Using default tag: latest
    The push refers to repository [192.168.1.231:5000/oa/labapp1]
    48cf90ddc02f: Pushed
    ded7a220bb05: Pushed
    latest: digest: sha256:97da93073218a7f1881ae5f21bdccdeccb93eccfd32579f09e74324197727b2d size: 735

18 $ tree /labregistry/
    /labregistry/
    └── docker
        └── registry
            └── v2
                ├── blobs
    ......
```

测试私有仓库中的镜像。仅保留必须的镜像 registry，清除实验环境中的其他所有镜像。示例命令如下：

```
19 $ docker rmi 192.168.1.231:5000/oa/labapp1:latest labapp1:latest alpine:latest

20 $ docker images
    REPOSITORY   TAG      IMAGE ID       CREATED        SIZE
    registry     latest   81c944c2288b   13 days ago    24.1MB
```

从私有仓库中下载、运行镜像，验证镜像中的文件，示例命令如下：

```
21 $ docker run 192.168.1.231:5000/oa/labapp1 cat /new.txt
    Unable to find image '192.168.1.231:5000/oa/labapp1:latest' locally
    latest: Pulling from oa/labapp1
    c158987b0551: Pull complete
    01b6bc151899: Pull complete
    Digest: sha256:97da93073218a7f1881ae5f21bdccdeccb93eccfd32579f09e74324197727b2d
    Status: Downloaded newer image for 192.168.1.231:5000/oa/labapp1:latest
    #下面是容器/new.txt 中的内容。
    Sat Nov 26 03:53:59 UTC 2022

22 $ docker images
    REPOSITORY                       TAG      IMAGE ID       CREATED        SIZE
    192.168.1.231:5000/oa/labapp1    latest   e1a9b9a51001   4 hours ago    7.05MB
    registry                         latest   81c944c2288b   13 days ago    24.1MB
```

8.11 习题

1. [单选题]以下关于 Docker 镜像描述，正确的是(　　)。

   A. Docker 镜像不是分层存储的架构

B. 构建镜像时，会一层层地构建。每一层构建完后，就不可以发生改变

C. 一个 Docker 镜像不可以作为其他镜像的基础镜像

D. Docker 镜像无法通过命令手动导入

2. [单选题]下列关于 Docker 镜像相关命令的执行，正确的是(　　)。

A. docker images 可查看现有镜像

B. docker push httpd 从 Docker Hub 拉取 httpd 镜像

C. docker load -o httpd 载入 httpd 镜像

D. docker rm httpd 删除 httpd 镜像

3. [单选题]Docker 镜像是一个(　　)的文件系统。

A. 可写　　B. 只读

C. 执行　　D. 完全控制

4. [单选题]使用 docker pull 命令下载镜像时，默认下载的镜像版本是(　　)。

A. latest　　B. older

C. small　　D. digest

5. [单选题]以下命令(　　)可以删除 docker 管理的所有本地镜像。

A. docker rmi 'docker images -q'　　B. docker rm 'docker images -q'

C. docker rm 'docker ps -aq'　　D. docker rmi 'docker ps -aq'

第9章

Docker容器管理

容器是镜像的运行实例，是提供对外服务的组件。Docker提供了十分丰富的操作命令用于高效地管理容器的整个生命周期。

本章要点

◎ 创建容器
◎ 停止或暂停容器
◎ 进入容器内部
◎ 删除容器
◎ 迁移容器
◎ 获得容器的更多信息

9.1 容器管理概述

如果容器没有运行，那么它仅仅是一个统一文件系统。如果容器运行了，它就是宿主机上的一个进程，与普通进程不同，它是一个隔离的进程空间。隔离进程空间中的进程可以创建、更改和删除文件，这些文件被保存在可读写的统一文件系统中，如图9-1所示。Docker使用了很多隔离技术，如果宿主机运行的是Linux操作系统，就会使用cgoup、名称空间等技术。

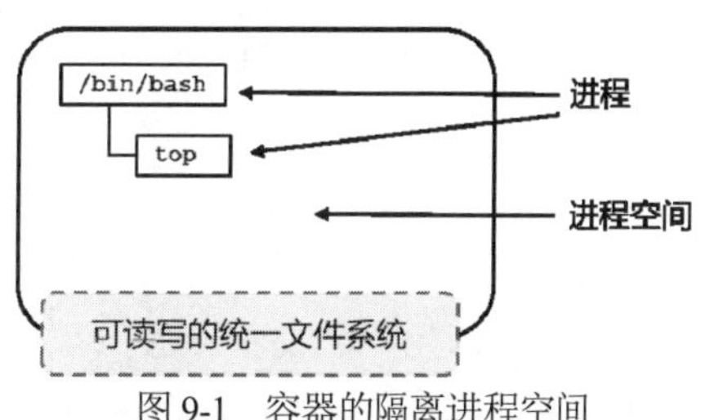

图9-1　容器的隔离进程空间

下面通过实验考察容器中的进程与可读写的统一文件系统，示例命令如下：

```
1 $ docker run -it ubuntu
    root@c21c5896f488:/# pwd
    /
    root@c21c5896f488:/# touch labtest1.txt
    root@c21c5896f488:/# exit
    exit

2 $ sudo find /var/lib/docker/overlay2/ -name labtest1.txt
  /var/lib/docker/overlay2/a99a567aefea22dad2c3086ca3fc59a70dfab9cdebed4bba33055e98334a20d4/diff/labtest1.txt
```

第 1 行命令在运行 Ubuntu 容器时使用了-it 选项，将进入容器中交互的 TTY。在容器的根目录下创建新的文件，文件会被保存在可读写的统一文件系统中。当退出这个容器时，它会自动停止。

第 2 行命令在宿主机中搜索新创建的文件，会在/var/lib/docker/overlay2 目录中容器的子目录中找到这个文件。

与宿主机中普通的进程类似，容器在其整个生命周期(图 9-2)中会处于不同的状态，共有以下 5 种状态：

- Created：已创建状态。
- Running：运行状态。
- Paused：已暂停状态。
- Stopped：已停止状态。
- Deleted：删除状态。

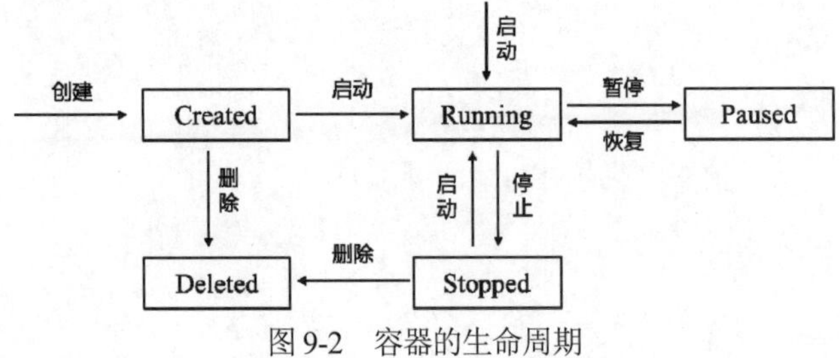

图 9-2　容器的生命周期

常用的容器操作子命令有：

- attach：将本地标准输入、输出和错误流附加到正在运行的容器。
- commit：根据容器创建新镜像。
- cp：在容器和宿主机之间复制文件或目录。
- create：创建新容器。
- diff：检查容器文件系统中文件或目录的更改。
- exec：在处于运行状态的容器中执行命令。
- export：将容器的文件系统导出为 TAR 格式的归档文件。
- inspect：显示一个或多个容器的详细信息。
- kill：杀死一个或多个处于运行状态的容器。
- logs：获取容器的日志。
- ls：列出容器。

- pause：暂停一个或多个容器中的所有进程。
- port：列出容器的端口映射或特定映射。
- prune：清除所有停止的容器。
- rename：重命名容器。
- restart：重新启动一个或多个容器。
- rm：移除一个或多个容器。
- run：在新容器中运行命令。
- start：启动一个或多个停止的容器。
- stats：实时显示容器资源的使用统计信息。
- stop：停止一个或多个处于运行状态的容器。
- top：显示容器中正在运行的进程。
- unpause：取消暂停一个或多个容器中的所有进程。
- update：更新一个或多个容器的配置。
- wait：阻止直到一个或多个容器停止，然后输出其退出代码。

9.2 创建容器

9.2.1 创建新容器

使用 create 子命令可以创建新容器。它的语法格式如下：

```
docker [container] create [OPTIONS] IMAGE [COMMAND] [ARG...]
```

create 子命令有很多选项，可以分为 3 类：

(1) 与容器运行模式相关的选项。

(2) 与容器环境配置相关的选项。

(3) 与容器资源限制安全保护相关的选项。

其中最常用的选项有：

- -i, --interactive：即使未连接，也保持 STDIN 打开。
- -t, --tty：分配伪 TTY。
- -h, --hostname：指定主机名。
- --name：指定容器的名称。

下面做一个创建容器的实验，示例命令如下：

```
1 $ docker create -it nonexistent
    Unable to find image 'nonexistent:latest' locally
    Error response from daemon: pull access denied for nonexistent, repository does not exist or may require
    'docker login': denied: requested access to the resource is denied

2 $ docker create -it ubuntu:latest
    Unable to find image 'ubuntu:latest' locally
```

```
latest: Pulling from library/ubuntu
125a6e411906: Pull complete
Digest: sha256:26c68657ccce2cb0a31b330cb0be2b5e108d467f641c62e13ab40cbec258c68d
Status: Downloaded newer image for ubuntu:latest
6ba84ed47e8b5825b8c2e4b31289478f8cc55ee2e96ae9779db3920b6c30d7dc

3 $ docker ps
CONTAINER ID   IMAGE   COMMAND   CREATED   STATUS   PORTS   NAMES

4 $ docker ps -a
CONTAINER ID   IMAGE          COMMAND   CREATED          STATUS    PORTS   NAMES
6ba84ed47e8b   ubuntu:latest  "bash"    38 seconds ago   Created           elastic_maxwell
```

创建容器时需要提供正确的镜像标识，例如：名称与 TAG，TAG 默认为 latest，否则就会像第 1 行命令的输出一样，提示有错误。

如果本地没有指定的镜像，就会像第 2 行命令一样自动下载镜像。下载成功后，会自动创建容器，最后一行会输出容器的长 ID。

有 3 种标识容器的方法：

- 长 ID：由 Docker 守护程序生成的 UUID。
- 短 ID：长 ID 字符串的前面 12 个字符。
- 名称：如果创建时没有通过--name 选项指定自定义的名称，则会自动生成一个随机名称。建议使用有意义的自定义名称，比 ID 容易记忆与描述。

提示

Docker 根据一个形容词列表和一个科学家姓名列表随机生成容器的名称。

为了方便自动化运维，可以在创建容器时通过--cidfile 选项将新容器的 ID 写入到指定的文件。

使用 ps、container ls 等多个子命令可以获得容器的列表，它们的语法格式如下：

```
docker [container] ps [OPTIONS]
docker container ls [OPTIONS]
docker container list [OPTIONS]
```

选项包括：

- -a, --all：显示所有容器，即使没有处于运行状态的容器。
- -f, --filter：根据条件过滤输出。
- --format：根据 Go 语言的模板格式化输出信息。
- -n, --last：显示 n 个上次创建的容器，默认值为-1。
- -l, --latest：显示最新创建的容器。
- --no-trunc：不截断输出。
- -q, --quiet：仅显示容器 ID。
- -s, --size：显示总文件大小。

输出信息包括以下字段：

- CONTAINER ID：容器的短 ID。
- IMAGE：镜像的名称与标签。
- COMMAND：运行容器后执行的命令。
- CREATED：容器的创建时间。
- STATUS：容器的状态。
- PORTS：端口。
- NAMES：容器的名称。

第 3 行命令仅显示正在运行的容器，所以在它的输出中没有容器的信息。

第 4 行命令中使用了-a 选项，所以会显示所有的容器。

从这个实验可以看出：通过 create 子命令创建的容器，并不会自动运行，还需要使用 start 子命令来启动它们。

9.2.2 启动容器

使用 start 子命令可以启动一个或多个已停止的容器。它的语法格式如下：

```
docker [container] start [OPTIONS] CONTAINER [CONTAINER...]
```

选项：

- -a, --attach：连接 STDOUT/STDERR 和转发信号。
- --detach-keys：替代用于分离容器的键序列。
- -i, --interactive：连接容器的 STDIN。

启动已停止的容器，示例命令如下：

```
1 $ docker create -it ubuntu
    64cbc775fcb01eda3891e0cd601768665976b7d000c15af2ff418f5e812d276c

2 $ docker ps -a
    CONTAINER ID   IMAGE     COMMAND   CREATED         STATUS    PORTS   NAMES
    64cbc775fcb0   ubuntu    "bash"    4 seconds ago   Created   determined_einstein

3 $ docker start 64cbc775fcb0
    64cbc775fcb0

4 $ docker ps -a
    CONTAINER ID IMAGE   COMMAND   CREATED          STATUS       PORTS   NAMES
    64cbc775fcb0 ubuntu  "bash"    43 seconds ago   Up 7 seconds  determined_einstein
```

通过第 4 行命令的输出可以看出：容器处于 Up 状态。

如果第 1 行命令在创建容器时，不使用-i 选项为新容器设置打开 STDIN，不使用-t 选项分配伪 TTY，那么容器启动完成之后会自动退出，容器将处于 Exited (0)状态，如下所示：

```
$ docker ps -a
CONTAINER ID   IMAGE           COMMAND   CREATED         STATUS                    PORTS   NAMES
768a727683fb   ubuntu:latest   "bash"    2 minutes ago   Exited (0) 1 second ago
                                                                           objective_meninsky
```

进入容器后可以执行各种命令。如果执行 exit 命令，将退出容器。容器退出之后将中止运行，状态更改为 exited。

如果需要临时断开与容器的连接，可以按下“分离键序列”，分离不影响容器的运行。默认的“分离键序列”是先按下 Ctrl+p 键，再按 Ctrl+q 键。执行 start 子命令时，可以使用--detach-keys 选项设置自定义的“分离键序列”，示例命令如下：

```
1 $ docker create -it ubuntu

2 $ docker ps -a
    CONTAINER ID  IMAGE  COMMAND   CREATED         STATUS    PORTS    NAMES
    532261053ef4  ubuntu  "bash"   3 seconds ago   Created   recursing_euclid

3 $ docker start -a -i --detach-keys "ctrl-x" 532261053ef4
    root@532261053ef4:/# top
# 如果按下 Ctrl+x 组合键，会断开与容器的会话，不影响容器的运行状态

4 $ docker ps -a
    CONTAINER ID  IMAGE  COMMAND   CREATED            STATUS          PORTS  NAMES
    532261053ef4  ubuntu  "bash"   About a minute ago  Up 33 seconds  recursing_euclid
```

提示

如果在容器中按下 Ctrl+c 组合键，通常会终止容器的运行。

9.2.3 新建并启动容器

除了“先使用 create 子命令创建容器，再通过 start 子命令启动容器”的方式之外，还可以直接使用 run 子命令创建并启动容器，它的工作原理如图 9-3 所示。当执行 run 子命令创建并启动容器时，Docker 在后台执行的操作包括：

- 检查宿主机是否存在指定的镜像。如果不存在，就从镜像仓库中下载。
- 利用镜像创建一个容器，并启动该容器。
- 在只读的镜像层上增加一个读写层。
- 从宿主机配置的网桥接口中桥接一个虚拟接口给容器。
- 从网桥的地址池中分配 IP 地址给容器。
- 执行用户指定的应用程序。
- 应用程序执行完毕后，容器被自动终止。

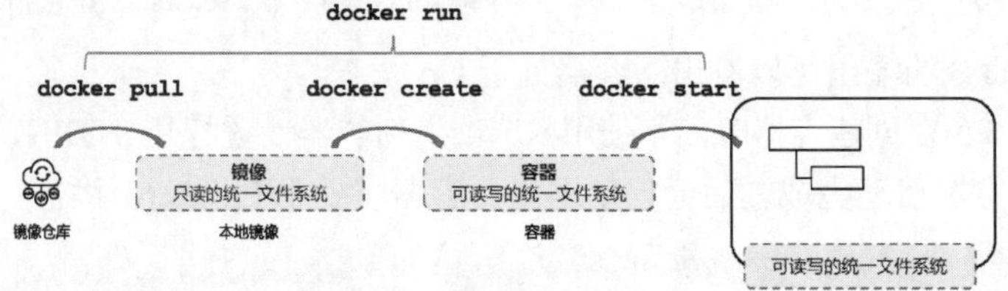

图 9-3 run 子命令的工作原理

可以把 run 子命令想象成一个“三合一”的子命令，它包含了 pull、create 和 start 三个子命

令的功能，所以与 create 和 start 子命令具有相同的选项，示例命令如下：

```
1 $ docker run busybox ping -c 2 8.8.8.8
    Unable to find image 'busybox:latest' locally
    latest: Pulling from library/busybox
    405fecb6a2fa: Pull complete
    Digest: sha256:fcd85228d7a25feb59f101ac3a955d27c80df4ad824d65f5757a954831450185
    Status: Downloaded newer image for busybox:latest
    PING 8.8.8.8 (8.8.8.8): 56 data bytes
    64 bytes from 8.8.8.8: seq=0 ttl=113 time=42.680 ms
    64 bytes from 8.8.8.8: seq=1 ttl=113 time=43.168 ms

    --- 8.8.8.8 ping statistics ---
    2 packets transmitted, 2 packets received, 0% packet loss
    round-trip min/avg/max = 42.680/42.924/43.168 ms

2 $ docker ps -a
    CONTAINER ID   IMAGE     COMMAND              CREATED          STATUS                PORTS   NAMES
  4f8966c7136f     busybox   "ping -c 2 8.8.8.8"  8 seconds ago    Exited (0) 7 seconds
                                                                  ago              sleepy_newton
```

从第 1 行命令的输出可以看出：与直接执行 ping 命令很类似，如果不是第一次执行时有一个下载镜像的操作，基本上感觉不出与普通的 ping 命令有什么区别。

提示

BusyBox 是一个组合了多种常见 UNIX/Linux 实用程序的工具集，被称为嵌入式 Linux 的“瑞士军刀”。

9.2.4 在后台运行容器

使用-d 或--detach 选项执行 run 子命令，会在后台运行容器并输出容器 ID。示例命令如下：

```
1 $ docker run -d httpd
    e80457bbc01e946f6ab9702d016be5831b01af71dc9d5531e93356f3aa0e6734

2 $ docker run -d alpine
    99a8b5b91f312b2bc09099c0789b3345eda08734fa7201ba698ebeadf0d4b6f7

3 $ docker run -d alpine /bin/sh -c "while true; do echo Hello; sleep 1; done"
    fd61fab2c5bb56c8a36d1af69a9f102d1c1d018b2794134dfea2a90b9f9431ee

4 $ docker ps -a
    CONTAINER ID IMAGE COMMAND          CREATED        STATUS        PORTS      NAMES
    99a8b5b91f31    alpine   "/bin/sh"             2 seconds ago   Exited (0) 1 second ago   quizzical_blackwell
    fd61fab2c5bb    alpine   "/bin/sh -c 'while…" About a minute ago Up About a   minute xenodochial_pasteur
    e80457bbc01e    httpd    "httpd-foreground" 2 minutes ago   Up 2 minutes   80/tcp    angry_babbage
```

第 1 行命令会在后台运行 HTTPD 容器。根据镜像的 CMD 属性的设置，会自动执行 httpd-foreground 脚本。由于这个脚本会持续运行，所以这个容器会持续在后台运行，从而达到类似守护程序(Daemon)的效果。

第 2 行命令在后台运行 Alpine 容器。根据镜像的 CMD 属性的设置，会自动执行/bin/sh。由于它不是可持续执行的命令或脚本，所以容器会退出。

第 3 行命令也是在后台运行 Alpine 容器。由于传递给容器的命令及参数构成一个死循环，所以容器不会退出，会持续地在后台运行。

提示

使用 inspect 子命令可以查看镜像的 CMD 属性。

9.2.5 查看容器输出

使用 logs 子命令可以查看日志的输出，它的语法格式如下：

```
docker [container] logs [OPTIONS] CONTAINER
```

选项：

- --details：输出详细信息。
- -f, --follow：持续保持输出。
- --since：输出从某个时间开始的日志。
- -n, --tail：输出最近的若干日志。
- -t, --timestamps：显示时间戳信息。
- --until：输出某个时间之前的日志。

查看在后台运行的容器的输出，示例命令如下：

```
1 $ docker run -d alpine /bin/sh -c "while true; do echo Hello; sleep 1; done"

2 $ docker ps -a
    CONTAINER ID   IMAGE    COMMAND                 CREATED         STATUS         PORTS   NAMES
    76879e76df18   alpine   "/bin/sh -c 'while t…"  7 seconds ago   Up 7 seconds   sad_driscoll

3 $ docker logs --details --follow --timestamps sad_driscoll
    2022-12-18T01:43:56.409673107Z   Hello
    2022-12-18T01:43:57.410862228Z   Hello
    2022-12-18T01:43:58.412593118Z   Hello
    2022-12-18T01:43:59.415657298Z   Hello
```

9.3 停止容器

9.3.1 暂停/恢复容器

使用 pause 子命令可以挂起指定容器中的所有进程，适用的场景有：

- 保证数据的一致性：在热备份或热迁移之前先暂停容器，待操作(例如创建卷的快照)

完成之后，再取消暂停、继续运行。

- 暂停业务应用：保存现有的状态，在需要时继续运行。

在 Linux 中的 Docker，实现暂停的方法是向容器的进程发送 freezer cgroup 信号，它比传统的 SIGSTOP 信号更可靠。

pause/unpause 子命令的语法格式如下所示：

```
docker [container] pause CONTAINER [CONTAINER...]
docker [container] unpause CONTAINER [CONTAINER...]
```

下面运行一个 Alpine Linux 容器，在进行操作时对其进行暂停操作，示例命令如下：

```
1 $ docker run -it --name test1 alpine
  / # vi
```

第 1 行命令通过--name 选项为容器设置名称，以方便后续的操作。在容器中运行 vi 编辑器，创建新文档，输入一些内容但不进行存盘操作，这些数据将被临时性地保存在容器的内存空间中。

再创建一个新的会话。在新的会话中执行后续的操作，示例命令如下：

```
2 $ docker ps
   CONTAINER ID    IMAGE   COMMAND    CREATED             STATUS               PORTS     NAMES
   fc3de42e090b    alpine  "/bin/sh"  About a minute ago  Up About a minute             test1

3 $ docker pause test1
   test1

4 $ docker ps -a
   CONTAINER ID   IMAGE    COMMAND    CREATED         STATUS                   PORTS   NAMES
   fc3de42e090b   alpine   "/bin/sh"  2 minutes ago   Up 2 minutes (Paused)    test1
```

第 4 行命令的输出显示：容器处于 Up 且 Paused 状态。这时对容器中的所有操作(例如在 vi 编辑操作)均无响应，容器就像被“冰封”了一样。

提示

虽然容器处于暂停状态，但是仍然可以通过 top、stats 等子命令获得容器的信息。

```
5 $ docker unpause test1
   test1

6 $ docker ps -a
   CONTAINER ID   IMAGE    COMMAND    CREATED         STATUS         PORTS   NAMES
   fc3de42e090b   alpine   "/bin/sh"  3 minutes ago   Up 3 minutes   test1
```

容器恢复后，可以立即进行后续的操作，例如：继续编辑文件，而且原来保存在内存中的数据没有丢失。

9.3.2 停止容器

使用 stop 子命令可以停止一个或多个正在运行的容器，它的语法格式如下所示：

```
docker [container] stop [OPTIONS] CONTAINER [CONTAINER...]
```

选项：

- -t，--time：在停止前等待停止的秒数(默认值为 10)。

默认情况下，stop 子命令向容器内的主进程将发送 SIGTERM(15)信号。容器的主进程如果能够对信号做出响应，执行正常的关闭操作，不会有数据丢失的风险，退出代码为 0。如果容器的主进程没有对 SIGTERM 信号做出响应，stop 子命令在等待一段宽限期(默认为 10 秒)后，会再发出 SIGKILL(9)信号强制关闭容器。强制关闭可能会导致据丢失，退出代码为 137。

关闭处于运行状态的 HTTPD 容器，示例命令如下：

```
1 $ docker run -d --name web1 httpd

2 $ docker ps -a
    CONTAINER ID    IMAGE    COMMAND               CREATED          STATUS          PORTS     NAMES
    c219630ae4ad    httpd    "httpd-foreground"    4 seconds ago    Up 3 seconds    80/tcp    web1

3 $ docker stop web1
    web1

4 $ docker ps -a
    CONTAINER ID    IMAGE    COMMAND               CREATED               STATUS                        PORTS    NAMES
    c219630ae4ad    httpd    "httpd-foreground"    About a minute ago    Exited (0) 19 seconds ago              web1
```

第 4 行命令的输出显示：容器 web1 处于退出状态，退出代码为 0，表示正常退出。如果通过 docker events 命令查看容器的实时事件，可以看到与退出相关的信息如下：

```
2022-12-28T11:29:32.357197277+08:00 container kill 容器 web1 的 ID (image=httpd, name=web1, signal=28)
2022-12-28T11:29:33.405967671+08:00 container die 容器 web1 的 ID (exitCode=0, image=httpd, name=web1)
2022-12-28T11:29:33.478667398+08:00 network disconnect 70d***c1c (container=容器 web1 的 ID,
  name=bridge, type=bridge)
2022-12-28T11:29:33.494604198+08:00 container stop 容器 web1 的 ID (image=httpd, name=web1)
```

Apache HTTPD 可以捕获 SIGWINCH(28)信号并进行正常的关闭。在 Docker 官方的 HTTP 镜像中，通过设置 StopSignal 属性对容器收到 SIGTERM 信号进行转换，转换为 SIGWINCH 信号。查看 HTTP 镜像 StopSignal 属性，示例如下：

```
5 $ docker inspect httpd | grep -i sign
            "StopSignal": "SIGWINCH"
            "StopSignal": "SIGWINCH"
```

关闭处于运行状态的 Alpine Linux 容器，示例命令如下：

```
6 $ docker run -it -d --name app1 alpine sh

7 $ time docker stop app1
    app1
```

```
real      0m10.190s
user      0m0.010s
sys       0m0.021s

8 $ docker ps -a
CONTAINER ID  IMAGECOMMAND          CREATED          STATUS                           PORTS   NAMES
68b08374bfb3  alpine  "sh"          2 minutes ago    Exited (137) About a minute ago          app1
c219630ae4ad  httpd   "httpd-foreground" 3 minutes ago Exited (0) 2 minutes ago              web1
```

第 7 行的命令输出显示：停止容器 app1 耗时 10 秒左右，这就是 stop 子命令在发出 SIGNTERM(15)信号后等待超时又发出 SIGNKILL(9)信号的时间，可以在 docker events 命令看到这个过程：

```
2022-12-28T11:30:40.403237643+08:00 container kill 容器 app1 的 ID (image=alpine, name=app1, signal=15)
2022-12-28T11:30:50.425246604+08:00 container kill 容器 app1 的 ID (image=alpine, name=app1, signal=9)
2022-12-28T11:30:50.463344016+08:00 container die 容器 app1 的 ID (exitCode=137, image=alpine,
  name=app1)
2022-12-28T11:30:50.527456422+08:00 network disconnect 70d***c1c (container=容器 app1 的 ID,
  name=bridge, type=bridge)
2022-12-28T11:30:50.538797963+08:00 container stop 容器 app1 的 ID (image=alpine, name=app1)
```

处于终止状态的容器，可以通过 start 子命令来重新启动。还可以通过 restart 子命令将一个正在运行的容器先终止，然后再重新启动。

9.3.3 杀死容器

使用 kill 子命令可以杀死一个或多个正在运行的容器。与 stop 子命令类似，也向容器内的主进程发送信号，不过默认的信号是 SIGKILL(9)。它的语法格式如下：

```
docker [container] kill [OPTIONS] CONTAINER [CONTAINER...]
```

选项：

- -s, --signal：送到容器的信号(默认为 KILL)。

可以通过如下命令查看 Linux 中的信号类型：

```
$ kill -l
 1) SIGHUP       2) SIGINT       3) SIGQUIT      4) SIGILL       5) SIGTRAP
 6) SIGABRT      7) SIGBUS       8) SIGFPE       9) SIGKILL     10) SIGUSR1
11) SIGSEGV     12) SIGUSR2     13) SIGPIPE     14) SIGALRM     15) SIGTERM
16) SIGSTKFLT   17) SIGCHLD     18) SIGCONT     19) SIGSTOP     20) SIGTSTP
21) SIGTTIN     22) SIGTTOU     23) SIGURG      24) SIGXCPU     25) SIGXFSZ
26) SIGVTALRM   27) SIGPROF     28) SIGWINCH    29) SIGIO       30) SIGPWR
31) SIGSYS      34) SIGRTMIN    35) SIGRTMIN+1  36) SIGRTMIN+2  37) SIGRTMIN+3
38) SIGRTMIN+4  39) SIGRTMIN+5  40) SIGRTMIN+6  41) SIGRTMIN+7  42) SIGRTMIN+8
43) SIGRTMIN+9  44) SIGRTMIN+10 45) SIGRTMIN+11 46) SIGRTMIN+12 47) SIGRTMIN+13
48) SIGRTMIN+14 49) SIGRTMIN+15 50) SIGRTMAX-14 51) SIGRTMAX-13 52) SIGRTMAX-12
53) SIGRTMAX-11 54) SIGRTMAX-10 55) SIGRTMAX-9  56) SIGRTMAX-8  57) SIGRTMAX-7
58) SIGRTMAX-6  59) SIGRTMAX-5  60) SIGRTMAX-4  61) SIGRTMAX-3  62) SIGRTMAX-2
63) SIGRTMAX-1  64) SIGRTMAX
```

杀死处于运行状态的 Alpine Linux 容器，示例命令如下：

```
1 $ docker run -it -d --name app2 alpine sh

2 $ time docker kill app2
    app2

    real      0m0.167s
    user      0m0.007s
    sys       0m0.024s

3 $ docker ps -a
    CONTAINER ID  MAGE  COMMAND CREATED         STATUS              PORTS    NAMES
    32f3fa201e34  lpine  sh"         23 seconds ago  Exited (137) 4 seconds ago        app2
```

可以在 docker events 命令看到这个过程：

```
2022-12-28T12:13:59.654614777+08:00 container kill  容器 app2 的 ID (image=alpine, name=app2, signal=9)
2022-12-28T12:13:59.692669191+08:00 container die  容器 app2 的 ID (exitCode=137, image=alpine,
    name=app2)
2022-12-28T12:13:59.759515958+08:00 network disconnect 70d4***c1c (container=容器 app2 的 ID,
    name=bridge, type=bridge)
```

9.4 进入容器内部

大多数情况下会使用-d 选项运行容器，令其在后台运行。但是为了排错、查看状态或执行其他命令，还需要进入容器内部。常见的方法是使用 attach 或 exec 子命令。

9.4.1 attach 子命令

使用 attach 子命令可以把本地标准输入、输出和错误流附加到正在运行的容器，可以把它简单地理解为连接到容器的控制台。它的语法格式如下：

```
docker [container] attach [OPTIONS] CONTAINER
```

选项：

- --detach-keys：设置退出会话的组合键。
- --no-stdin：关闭标准输入。
- --sig-proxy：转发收到的系统信号给应用进程，默认为 true。

进入在后台运行的 Alpine Linux 容器，示例命令如下：

```
1 $ docker run -itd --name test1 alpine
    8a19980c37e518bfaa17b2c0bbb183778efcff5277bbe73e59a95ca0431c09f5

2 $ docker ps
    CONTAINER ID    IMAGE    COMMAND       CREATED          STATUS         PORTS   NAMES
    8a19980c37e5    alpine   "/bin/sh"     8 seconds ago    Up 6 seconds           test1

3 $ docker attach test1
```

```
    / # hostname
    8a19980c37e5     按下 Ctrl+p Ctrl+q 组合键退出
    / # read escape sequence

4 # docker attach --detach-keys "ctrl-d" rtest1
    / # hostname
    8a19980c37e5     按下 Ctrl+d 组合键退出
    / # read escape sequence

5 $ docker ps
    CONTAINER ID  IMAGE   COMMAND    CREATED          STATUS          PORTS    NAMES
    8a19980c37e5  alpine  "/bin/sh"  30 seconds ago   Up 27 seconds            test1
```

第 1 行命令使用-itd 选项在后台运行容器。这个容器会有一个用于交互的控制台。

第 3 行命令连接到这个交互的控制台，可以在其中执行操作。操作完成后，要注意退出的方式：

- 输入 exit 命令或者按下 Ctrl+d 组合键，会导致容器的中止退出。
- 使用 Ctrl+p、Ctrl+q 组合键，仅仅是断开控制台，容器还会在后台运行。
- 如果容器中的 Ctrl+p、Ctrl+q 组合键被占用，则可以像第 4 行命令一样通过--detach-keys 来指定其他组合键。

第 5 行命令的输出显示：由于采用的是断开的方式，所以容器还处于运行状态。

如果同时在多个会话使用 attach 子命令连接到同一个容器，那么所有会话将同步显示相同的内容。如果某个会话中执行的操作被阻塞，那么其他会话也无法进行操作。

9.4.2 exec 子命令

使用 exec 子命令可以在一个正在运行的容器中运行命令。与 attach 子命令相比，它是“自起炉灶”，不会附加到现有的容器控制台，而且命令执行结束会退出，不会影响容器在后台执行。它的语法格式如下：

```
docker [container] exec [OPTIONS] CONTAINER COMMAND [ARG...]
```

选项：

- -d, --detach：分离模式，在后台运行命令。
- --detach-keys：替代用于分离容器的键序列。
- -e, --env：设置环境变量。
- --env-file：读取环境变量文件。
- -i, --interactive：即使未连接，也保持 STDIN 打开。
- --privileged：为命令授予扩展权限。
- -t, --tty：分配伪 TTY。
- -u, --user：用户名或 UID(格式：<name|uid>[:<group|gid>])。
- -w, --workdir：容器内的工作目录。

进入在后台运行的 Alpine Linux 容器，示例命令如下：

```
1 $ docker run -itd --name test2 alpine
```

```
2 $ docker ps -a
    CONTAINER ID  IMAGE   COMMAND    CREATED        STATUS        PORTS   NAMES
    133e4b6cf46b  alpine  "/bin/sh"  4 seconds ago  Up 4 seconds          test2

3 $ docker exec -it test2 sh
    / # hostname
    133e4b6cf46b
    / # exit

4 $ docker ps -a
    CONTAINER ID  IMAGE   COMMAND    CREATED         STATUS         PORTS  NAMES
    133e4b6cf46b  alpine  "/bin/sh"  25 seconds ago  Up 24 seconds         test2
```

第 3 行命令通过指定-it 选项保持标准输入，并且分配一个伪终端，在容器中执行 sh 命令会创建一个新的 Shell 环境，在不影响容器内其他应用的前提下，用户可以与容器进行交互操作。

相对于 attach 子命令，通过 exec 子命令对容器执行操作是推荐的方式。

9.5 删除容器

9.5.1 rm 子命令

使用 rm 子命令可以删除一个或多个容器。它的语法格式如下：

```
docker container rm [OPTIONS] CONTAINER [CONTAINER...]
```

选项：

- -f, --force：强制删除正在运行的容器(使用 SIGKILL 信号)。
- -l, --link：删除指定的链接。
- -v, --volumes：删除与容器关联的匿名卷。

删除处于不同状态的容器，示例命令如下：

```
1 # docker run -itd --name test1 alpine

2 # docker run    --name test2 ubuntu

3 # docker ps -a
    CONTAINER ID  IMAGECOMMAND       CREATED         STATUS                    PORTS   NAMES
    c4679f1fa85a  ubuntu  "bash"     6 seconds ago   Exited (0) 4 seconds ago          test2
    f3c8fc45a3bb  alpine  "/bin/sh"  36 seconds ago  Up 35 seconds                     test1
```

第 3 行命令输出显示：容器 test1 处于运行状态，容器 test2 处于终止状态。

```
4 # docker rm test1
    Error response from daemon: You cannot remove a running container
        f3c8fc45a3bbc7451377b11874dd439af788f4ef0aa67b75ca3ffed458dd0fd8. Stop the container before
        attempting removal or force remove
```

```
5 # docker rm --force test1
  test1
```

rm 子命令默认只能删除处于终止状态的容器，所以第 4 行命令会出错，必须使用-f 或--force 选项进行强制删除。Docker 会先发送 SIGKILL 信号给容器，待终止之后再进行删除。

```
6 # docker rm $(docker ps --filter status=exited -q)
  c4679f1fa85a

7 # docker ps -a
  CONTAINER ID   IMAGE   COMMAND   CREATED   STATUS   PORTS   NAMES
```

第 6 行命令先使用 docker ps --filter status=exited -q 获得要删除的容器 ID 列表，并将其作为 rm 子命令的参数，这样可以移除所有停止的容器。

除了使用 rm 子命令手工删除容器之外，还可以在执行 run 子命令时使用--rm 选项，从而在容器退出后自动删除容器。示例命令如下：

```
1 $ docker run -itd --name test1 --rm alpine sh

2 $ docker ps -a
  CONTAINER ID   IMAGE    COMMAND   CREATED         STATUS       PORTS   NAMES
  816dcb759caa   alpine   "sh"      3 seconds ago   Up 2 seconds         test1

3 $ docker stop test1

4 $ docker ps -a
  CONTAINER ID   IMAGE   COMMAND   CREATED   STATUS   PORTS   NAMES
```

9.5.2 prune 子命令

使用 prune 子命令可以删除所有已停止的容器。它的语法格式如下：

```
docker container prune [OPTIONS]
```

选项：

- --filter：设置过滤器，例如：'until=<timestamp>'。
- -f, --force：不需要确认。

如果要删除 2022-12-27 13:10:00 之前创建的容器，则示例命令如下：

```
1 $ docker ps -a --format 'table {{.ID}}\t{{.Image}}\t{{.Command}}\t{{.CreatedAt}}\t{{.Status}}'
  CONTAINER ID   IMAGE     COMMAND   CREATED AT                      STATUS
  53a9bc23a516   busybox   "sh"      2022-12-27 13:11:59 -0800 PST   Exited (0) 17 minutes ago
  4a75091a6d61   busybox   "sh"      2022-12-27 13:09:53 -0800 PST   Exited (0) 19 minutes ago

2 $ docker container prune --force --filter "until=2022-12-27T13:10:00"
  Deleted Containers:
  4a75091a6d618526fcd8b33ccd6e5928ca2a64415466f768a6180004b0c72c6c
  Total reclaimed space: 27 B

3 $ docker ps -a --format 'table {{.ID}}\t{{.Image}}\t{{.Command}}\t{{.CreatedAt}}\t{{.Status}}'
  CONTAINER ID   IMAGE   COMMAND   CREATED AT   STATUS
```

```
53a9bc23a516        busybox    "sh"      2022-12-27 13:11:59 -0800 PST    Exited (0) 19 minutes ago
```

第 1 行命令通过 Go 语言格式的模板控制输出信息(\t 在字段之间添加制表符)。

第 2 行命令仅清除符合过滤条件的容器。

9.6 迁移容器

不依赖任何第三方工具，仅使用 export 与 import 子命就可以实现容器的迁移。在源宿主机上，使用 export 子命令将容器的导出存为 TAR 格式的归档文件；在目标宿主机上，通过 import 子命令导入文件构建新镜像，然后使用镜像再创建新的容器，这样就实现了容器的迁移。

不管容器处于何种状态，都可以使用 export 进行导出。它的语法格式如下：

```
docker [container] export [OPTIONS] CONTAINER
```

选项：

- -o, --output：导出到文件而不是标准输出。

导出一个名为 test1 的容器，示例命令如下：

```
1 $ docker export test1 -o export.tar

2 $ mkdir exptest

3 $ tar -xf export.tar -C exptest/

4 $ tree -L 1 exptest/
    exptest/
    ├── bin
    ├── dev
    ├── etc
    ├── home
    ├── lib
    ├── media
    ├── mnt
    ├── opt
    ……略……
```

第 1 行命令导出的文件是 export.tar。不管容器与镜像有多少层，export 操作的对象是合并后的统一文件系统，所以导出的文件中既没有分层结构也没有对应的元数据。

第 3 行命令将 export.tar 文件展开。从第 4 行命令的输出可以看出：这是一个单一的目录树结构。

使用 export 子命令进行导入。导入的结果是创建一个只有一层的镜像，它的语法格式如下：

```
docker [image] import [OPTIONS] file|URL|- [REPOSITORY[:TAG]]
```

选项：

- -c, --change：为创建出的镜像设置 Dockerfile 格式的指令。
- -m, --message：注释信息。
- --platform：设置平台的信息。

在目标宿主机上进行导入操作，示例命令如下：

```
1 $ docker import export.tar myweb1
    sha256:1575863e494714d7c079425dafdfca333795ba1b6ab37fc805fd744ee87430f3

2 $ docker images
    REPOSITORY    TAG       IMAGE ID        CREATED          SIZE
    myweb1        latest    1575863e4947    4 seconds ago    5.53MB

3 $ docker run -itd myweb1
```

9.7 查看容器

9.7.1 查看容器详情

inspect 是一个通用子命令，可以查看多种 Docker 对象的底层信息。它的语法格式如下：

```
docker inspect [OPTIONS] NAME|ID [NAME|ID...]
```

选项：

- --format：使用 Go 语言模板控制输出信息。
- -s, --size：显示总文件大小。
- --type string：返回指定类型的 JSON 格式的信息。

如果仅仅查看容器的详细信息，可以使用 container inspect 子命令，它的语法格式如下：

```
docker container inspect [OPTIONS] CONTAINER [CONTAINER...]
```

选项：

- --format：使用 Go 语言模板控制输出信息。
- -s, --size：显示总文件大小。

查看指定容器的信息，示例命令如下：

```
1 $ docker container inspect 92adb3a5849d
    [
        {
            "Id": "92adb3a5849dcbb4ad21896fd46b836870f0d170fe13aa66506c75844a6bcd63",
            "Created": "2022-05-23T13:57:19.514426489Z",
            "Path": "sh",
            "Args": [],
            "State": {
                "Status": "exited",
```

```
                    "Running": false,
                    "Paused": false,
    ……
2 $ docker container inspect -f '{{range .NetworkSettings.Networks}}{{.IPAddress}}{{end}}' mynginx1
    172.17.0.2
```

第 1 行命令输出的容器详细信息是 JSON 格式的，包括容器 ID、创建时间、文件路径、状态、镜像、网络地址等属性。

如果仅关注某些信息，可以通过-f 来设置输出模板，达到类似过滤的效果，例如第 2 行命令会只输出容器的 IP 地址。

9.7.2 查看容器内进程

类似于 Linux 系统中的 top 命令，可以在 Docker 中使用 top 子命令查看处于运行状态的容器中的进程信息。它的语法格式如下：

```
docker container top CONTAINER [ps OPTIONS]
```

top 子命令使用 ps 子命令的选项，输出的信息包括 PID、用户、时间和命令等。

查看指定容器内的进程信息，示例命令如下：

```
1 $ docker run -d ubuntu /bin/bash -c "while true; do echo hello; sleep 2; done"

2 $ docker ps
    CONTAINER ID    IMAGE    COMMAND                   CREATED          STATUS          PORTS    NAMES
    98a10b4b2a28    ubuntu   "/bin/bash -c 'while…"    27 seconds ago   Up 26 seconds            bold_mclaren

3 $ docker top bold_mclaren
    UID     PID     PPID    C    STIME    TTY    TIME        CMD
    root    2561    2533    0    08:08    ?      00:00:00    /bin/bash -c while true 略……
    root    2618    2561    0    08:08    ?      00:00:00    sleep 2
```

9.7.3 查看统计信息

利用 stats 子命令可以查看容器的 CPU、内存、存储和网络等资源使用情况。它的语法格式如下：

```
docker container stats [OPTIONS] [CONTAINER...]
```

选项：

- -a, --all：输出所有容器的统计信息，默认只输出处于运行状态的容器。
- --format：格式化输出信息。
- --no-stream：不持续输出，默认会自动更新持续实时结果。
- --no-trunc：不截断输出信息。

查看指定容器内的进程信息，示例命令如下：

```
$ docker stats test1 --no-stream
  CONTAINER ID  NAME          CPU %   MEM USAGE / LIMIT    MEM %   NET I/O       BLOCK I/O     PIDS
  98a10b4b2a28  bold_mclaren  1.00%   2.391MiB / 7.479GiB  0.02%   1.16kB / 0B   73.7kB / 0B   2
```

9.8 其他容器命令

9.8.1 复制文件

使用 cp 子命令可以在容器与宿主机之间复制文件或目录。它的语法格式如下：

```
docker [container] cp [OPTIONS] CONTAINER:SRC_PATH DEST_PATH|-
docker cp [OPTIONS] SRC_PATH|- CONTAINER:DEST_PATH
```

如果使用“-”作为源，则会从 STDIN 读取 TAR 存档，并将其提取到容器中的目标目录。如果使用“-”作为目标，则会将容器源的 TAR 存档输出到 STDOUT。

选项：

- -a, --archive：打包模式，复制文件会带有原始的 UID/GID 信息。
- -l, --follow-link：跟随软链接。当原路径为软链接时，默认只复制链接信息。

将宿主机当前目录下的文件复制到容器的根目录下，示例命令如下：

```
1 $ docker run -itd --name test1 alpine sh

2 $ docker ps -a
    CONTAINER ID   IMAGE   COMMAND   CREATED          STATUS         PORTS   NAMES
    867422e5102c   alpine  "sh"      15 seconds ago   Up 14 seconds          test1

3 $ docker cp 1.txt test1:/

4 $ docker exec test1 ls -l /1.txt
    -rw-r--r--     1 1000     1000           32 Dec 28 23:36 /1.txt
```

9.8.2 查看变更

使用 diff 子命令可以查询保存在容器读写层中的信息，并输出容器内文件系统的变更。它的语法格式如下：

```
docker [container] diff CONTAINER
```

查看基于 HTTPD 镜像创建的容器文件的变化，示例命令如下：

```
1 $ docker run -itd --name web1 httpd

2 $ docker cp 1.txt web1:/

3 $ docker diff web1
    C /usr
    C /usr/local
    C /usr/local/apache2
    C /usr/local/apache2/logs
    A /usr/local/apache2/logs/httpd.pid
```

```
A /1.txt
```

diff 子命令列出容器自创建以来发生过更改的文件和目录。有三种不同类型的变更，如表 9-1 所示。

表 9-1　diff 子命令输出信息对照表

符号	说明
A	已添加文件或目录
D	文件或目录已删除
C	文件或目录已更改

9.8.3　查看端口映射

使用 port 子命令可以列出容器的端口映射或特定映射。它的语法格式如下：

```
docker [container] port CONTAINER [PRIVATE_PORT[/PROTO]]
```

查看容器 web1 的端口映射情况，示例命令如下：

```
1 $ docker run -d -p 80:80 --name web1 httpd

2 # docker ps
    CONTAINER ID   IMAGE   COMMAND   CREATED   STATUS   PORTS   NAMES
    467497f8551b   httpd "   httpd-foreground"   2 minutes ago Up   2 minutes   0.0.0.0:80->80/tcp, :::80->80/tcp   web1

3 # docker container port web1
    80/tcp -> 0.0.0.0:80
    80/tcp -> :::80

4 $ curl localhost
    <html><body><h1>It works!</h1></body></html>

5 $ docker container port web1 8080
    Error: No public port '8080/tcp' published for web1
```

第 1 行命令使用-p 选项将容器的 80 端口映射到宿主机的 80 端口。第 2 行、第 3 行命令的输出显示了这种映射关系。

如果查看的容器没有映射端口，则会出现类似第 5 行命令输出的错误。

9.8.4　更新配置

使用 update 子命令可以更新容器的一些运行时配置，主要是一些资源限制份额。它的语法格式如下：

```
docker [container] update [OPTIONS] CONTAINER [CONTAINER .. . ]
```

选项：

- --blkio-weight：块 IO 的权重。

- --cpu-period：限制 CPU 完全公平调度器(completely fair scheduler，CFS)的使用时间。
- --cpu-quota：限制 CPU 调度器的配额。
- --cpu-rt-period：限制 CPU 调度器的实时周期，单位为微秒。
- --cpu-rt-runtime：限制 CPU 调度器的实时运行时间，单位为微秒。
- -c, --cpu-shares：限制 CPU 的使用份额。
- --cpus：限制 CPU 的数量。
- --cpuset-cpus：允许使用的 CPU 核，如 0-3,0,1。
- --cpuset-mems：允许使用的内存块，如 0-3,0,1。
- --kernel-memory：限制使用的内核内存。
- -m, --memory：限制使用的内存。
- --memory-reservation：内存软限制。
- --memory-swap：内存加上缓存区的限制，－1 表示对缓冲区无限制。
- --pids-limit：调整容器 pids 的限制，－1 表示无限制。
- --restart：容器退出后的重启策略。

更新容器的 CPU 共享设置，示例如下：

```
1 $ docker update --cpu-shares 512 test1

2 $ docker update --cpu-shares 512 -m 300M qazde7571666 test1
```

第 1 行命令将容器的 CPU 共享限制为 512。

第 2 行命令将更新多个容器的多个资源配置。

9.9 习题

1. [单选题]下列关于 Docker 容器，说法错误的是(　　)。
   A. 镜像是只读模板，容器是给这个只读模板添加额外的可写层
   B. 容器是轻量级，用户可以随时创建或删除
   C. 通过 docker create 命令创建的容器，默认是启动状态的
   D. 容器是与其中运行的 shell 命令共存亡的终端，命令运行容器运行，命令结束容器退出
2. [多选题]docker run 命令是(　　)与(　　)命令的结合体。
   A. docker create　　B. docker start
   C. docker stop　　D. docker rm
3. [单选题]查看容器的详细信息的命令是(　　)。
   A. docker ps　　B. docker inspect
   C. docker stats　　D. docker logs
4. [单选题]下列(　　)命令能查看到已经停止的容器。
   A. docker ps　　B. docker ps -a
   C. docker container ls　　D. docker container ls -a

5. [单选题]选项(　　)能创建一个 nginx 容器并放到后台运行。

A. docker run -d nginx　　B. docker run -it nginx

C. docker run nginx　　D. docker -i nginx

6. [单选题]选项(　　)能进入容器终端。

A. docker exec -d nginx　　B. docker exec -it nginx /bin/bash

C. docker exec -itd nginx　　D. docker exec -h nginx

7. [单选题]下列容器的相关命令中，说法错误的是(　　)。

A. docker start 28edb150112c 启动 ID 为 28edb150112c 的容器

B. docker exec -it 28edb150112c /bin/bash 进入 ID 为 28edb150112c 的容器

C. docker export 28edb150112c -o newcontainer.tar 将容器 28edb150112c 导出生成 newcontainer.tar

D. docker rmi 28edb150112c 删除 ID 为 28edb150112c 的容器

8. [单选题]若要将容器强制删除，需要在 docker rm 命令中添加的参数是(　　)。

A. –f　　B. --no-trunc

C. –q　　D. -a

9. [单选题]当 Docker 容器遭遇黑客攻击时，应该使用命令(　　)保存容器的当前状态为镜像。

A. docker commit　　B. docker save

C. docker export　　D. docker diff

10. [单选题]选项(　　)将创建一个 Nginx 容器并暴露到宿主机 80 端口。

A. docker run -d 80 nginx

B. docker run -d -p 80:80 nginx

C. docker run -port 80:80 nginx

D. docker run --sport 80 --dport 80 nginx

第 10 章

Docker网络管理

Docker 容器对外提供服务时通常需要网络，可以通过网络隔离来保护容器。本章将学习容器与容器之间、容器与宿主机之间及与外部网络之间的通信。

本章要点

◎ Docker 网络的启动与配置

◎ 容器的名称解析

◎ 容器的访问控制

◎ 容器的端口映射

◎ 容器的便捷互联机制

◎ 容器的网络管理命令

10.1 Docker 网络的启动和配置

10.1.1 网络启动过程

Docker 守护程序启动时，首先会在宿主机上自动创建一个名为 docker0 的虚拟网桥，可以把虚拟网桥理解为一个网络交换机，负责在挂载其上的接口之间进行包转发。然后随机分配一个本地未占用的私有网段(在 RFCl918 中定义)中的 IPv4 地址给 docker0 接口，例如：默认是 172.17.0.0/16 网段，将第 1 个地址给 docker0，如图 10-1 所示。

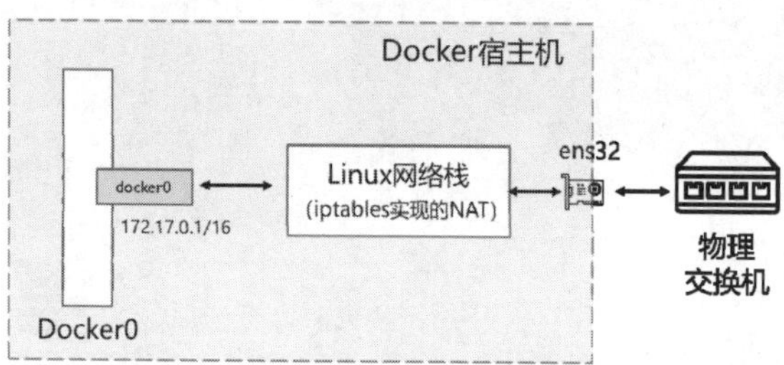

图 10-1　未运行容器之前的宿主机网络

可以从系统日志中查看相应的信息，示例命令如下：

```
$ sudo tail -f /var/log/messages
……
Jul 9 20:55:08 docker1 dockerd[2959]: time="2022-07-09T20:55:08.322682227+08:00" level=info msg="Default
   bridge (docker0) is assigned with an IP address 172.17.0.0/16. Daemon option --bip can be used to set a
   preferred IP address"
……
```

通过 ip address 命令查看 docker0 的信息，示例命令如下：

```
$ ip address show docker0
   3: docker0: <NO-CARRIER,BROADCAST,MULTICAST,UP> mtu 1500 qdisc noqueue state DOWN group
     default
     link/ether 02:42:51:8c:66:bb brd ff:ff:ff:ff:ff:ff
     inet 172.17.0.1/16 brd 172.17.255.255 scope global docker0
            valid_lft forever preferred_lft forever
```

当运行容器之后，默认会增加 1 个虚拟以太网对(veth pair)，一端是容器的 eth0，另外一端是网桥 docker0 的接口 vethID，如图 10-2 所示。

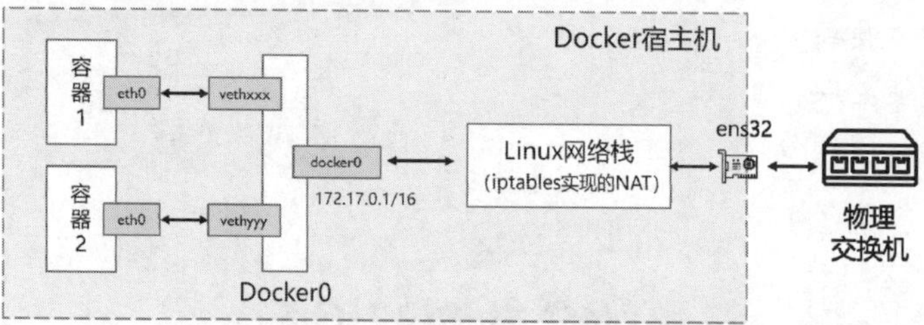

图 10-2　运行容器之后的宿主机网络

启动第 1 个容器，考察网络的变化，示例命令如下：

```
1 $ docker run -itd --name c1 --rm alpine sh

2 $ sudo tail -f /var/log/messages
     ……
     Jul 19 15:40:49 docker1 kernel: eth0: renamed from veth4a62fd8
     Jul 19 15:40:49 docker1 kernel: IPv6: ADDRCONF(NETDEV_CHANGE): vethc753b9c: link becomes
       ready
     Jul 19 15:40:49 docker1 kernel: docker0: port 1(vethc753b9c) entered blocking state
     Jul 19 15:40:49 docker1 kernel: docker0: port 1(vethc753b9c) entered forwarding state
     Jul 19 15:40:49 docker1 NetworkManager[814]: <info>   [1658216449.4150] device (vethc753b9c): carrier:
       link connected
```

```
    Jul 19 15:40:49 docker1 NetworkManager[814]: <info>  [1658216449.4160] device (docker0): carrier: link
      connected
    ......

3 $ ip address
    ......
    3: docker0: <BROADCAST,MULTICAST,UP,LOWER_UP> mtu 1500 qdisc noqueue state UP group
      default
        link/ether 02:42:49:fc:07:c2 brd ff:ff:ff:ff:ff:ff
        inet 172.17.0.1/16 brd 172.17.255.255 scope global docker0
            valid_lft forever preferred_lft forever
        inet6 fe80::42:49ff:fefc:7c2/64 scope link
            valid_lft forever preferred_lft forever
    11: vethc753b9c@if10: <BROADCAST,MULTICAST,UP,LOWER_UP> mtu 1500 qdisc noqueue master
      docker0 state UP group default
        link/ether ae:54:74:8f:ee:ff brd ff:ff:ff:ff:ff:ff link-netnsid 0
        inet6 fe80::ac54:74ff:fe8f:eeff/64 scope link
        valid_lft forever preferred_lft forever
```

第 2 行、第 3 行命令的输出显示：宿主机新增加 1 个名为 vethc753b9c@if10 的网络接口，显示的属性是 master docker0，说明它是网桥 docker0 的一个端口。

再启动第 2 个容器，考察网络的变化，示例命令如下：

```
4 $ docker run -itd --name c2 --rm alpine sh

5 $ sudo tail -f /var/log/messages
    ......
Jul 19 15:45:02 docker1 systemd[1]: Started libcontainer container
  b005deb481821a091e76b7bc2c2a75d8795d544ead8059a24bc2fa58fd82666d.
Jul 19 15:45:02 docker1 kernel: eth0: renamed from vetha6ef22e
Jul 19 15:45:02 docker1 kernel: IPv6: ADDRCONF(NETDEV_CHANGE): vethefe082c: link becomes ready
Jul 19 15:45:02 docker1 kernel: docker0: port 2(vethefe082c) entered blocking state
Jul 19 15:45:02 docker1 kernel: docker0: port 2(vethefe082c) entered forwarding state
Jul 19 15:45:02 docker1 NetworkManager[814]: <info>  [1658216702.4982] device (vethefe082c): carrier: link
    connected
    ......
```

第 5 行命令的输出显示：宿主机又新增了 1 个虚拟以太网对。

10.1.2 网络配置参数

查看 Docker 守护程序的状态，示例命令如下：

```
$ sudo systemctl status docker
● docker.service - Docker Application Container Engine
    Loaded: loaded (/usr/lib/systemd/system/docker.service; disabled; vendor preset: disabled)
    Active: active (running) since Tue 2022-07-19 15:26:35 CST; 21min ago
    TriggeredBy: ● docker.socket
    Docs: https://docs.docker.com
    Main PID: 4457 (dockerd)
    Tasks: 9
```

```
    Memory: 133.7M
       CPU: 2.073s
    CGroup: /system.slice/docker.service
            └─4457 /usr/bin/dockerd -H fd:// --containerd=/run/containerd/containerd.sock
```

查看 Docker 守护程序的配置文件的目录，示例命令如下：

```
$ ls -l /etc/docker/
total 4
-rw-------. 1 root root 244 Jul 28 11:37 key.json
```

由于当前环境中没有配置文件，因此将使用默认的参数启动 Docker 守护程序。查看 dockerd 命令的参数，示例命令如下：

```
$ dockerd –help
```

常用的参数选项有：

- --bip：设置网桥的 IP 地址。
- -b, --bridge：设置容器挂载的网桥。
- -H, --host：连接到的 Docker 服务端。
- --icc：启用容器内通信(inter container communication)功能。
- --ip-forward：启用 net.ipv4_forward，即打开转发功能。
- --iptables：启用 iptables 规则。
- --mtu：设置容器网络中的最大传输单元(MTU)。
- --dns：设置 DNS 服务器。
- --dns-option：设置 DNS 选项。
- --dns-search：设置 DNS 搜索域。

10.2 容器的名称解析

10.2.1 名称解析器默认的配置

由于通常会给容器分配动态 IP 地址，所以为了实现访问的一致性，建议使用主机名或域名来访问。考察容器默认的主机名及解析器的设置，示例命令如下：

```
$ docker run -it alpine sh

  / # hostname
  f44aa4e194ee

  / # ip addr
  1: lo: <LOOPBACK,UP,LOWER_UP> mtu 65536 qdisc noqueue state UNKNOWN qlen 1000
      link/loopback 00:00:00:00:00:00 brd 00:00:00:00:00:00
      inet 127.0.0.1/8 scope host lo
         valid_lft forever preferred_lft forever
```

```
4: eth0@if5: <BROADCAST,MULTICAST,UP,LOWER_UP,M-DOWN> mtu 1500 qdisc noqueue state UP
    link/ether 02:42:ac:11:00:02 brd ff:ff:ff:ff:ff:ff
    inet 172.17.0.2/16 brd 172.17.255.255 scope global eth0
       valid_lft forever preferred_lft forever

/ # mount | grep etc
/dev/mapper/cs-root on /etc/resolv.conf type xfs
  (rw,seclabel,relatime,attr2,inode64,logbufs=8,logbsize=32k,noquota)
/dev/mapper/cs-root on /etc/hostname type xfs
  (rw,seclabel,relatime,attr2,inode64,logbufs=8,logbsize=32k,noquota)
/dev/mapper/cs-root on /etc/hosts type xfs (rw,seclabel,relatime,attr2,inode64,logbufs=8,logbsize=32k,noquota)
```

Docker 默认会为容器分配随机生成的主机名。与容器的主机名和 DNS 服务器相关的配置文件有/etc/resolv.conf、/etc/hosts 和/etc/hostname，在容器中使用 mount 命令可以看到与这 3 个文件有关的挂载信息。

Docker 启动容器时，会复制宿主机的/etc/resolv.conf 文件的设置，并删除掉其中无法连接的 DNS 服务器，示例命令如下：

```
/ # cat /etc/resolv.conf
# Generated by NetworkManager
  nameserver 114.114.114.114
```

查看容器的主机名称，示例命令如下：

```
/ # cat /etc/hostname
f44aa4e194ee
```

查看容器的 hosts 文件，示例命令如下：

```
/ # cat /etc/hosts
127.0.0.1 localhost
::1 localhost ip6-localhost ip6-loopback
fe00::0 ip6-localnet
ff00::0 ip6-mcastprefix
ff02::1 ip6-allnodes
ff02::2 ip6-allrouters
172.17.0.2 f44aa4e194ee
```

/etc/hosts 文件默认只记录了容器自身的地址和主机名称的映射。

10.2.2　修改解析器的配置

在容器运行时，虽然可以直接编辑容器中的/etc/hosts、/etc/hostname 和/etc/resolve.conf 等配置文件，但是这些修改都是临时的，当容器停止或重新启动之后将被丢弃，同时也无法被 commit 子命令提交新镜像中。所以，如果需要修改解析器的配置，最佳的策略是在运行容器时通过参数来指定：

- --dns：设置 DNS 服务器。
- --dns-option：设置 DNS 选项。
- --dns-search：设置 DNS 搜索域。
- -h, --hostname：设置容器的主机名。

- --ip：设置容器的 IPv4 地址。
- --link：添加指向另一个容器的链接。

通过这些选项运行容器，会自动修改容器中与名称解析有关的配置，示例命令如下：

```
1 $ cat /etc/resolv.conf
    # Generated by NetworkManager
    nameserver 8.8.8.8
    nameserver 114.114.114.114

2 $ docker run -itd --rm --name db1 alpine sleep 1d

3 $ docker ps -a
    CONTAINER ID    IMAGE    COMMAND      CREATED          STATUS         PORTS    NAMES
    29a5bbcbc24e    alpine   "sleep 1d"   6 seconds ago    Up 4 seconds   db1
```

第 1 行命令显示宿主机的解析器设置。

第 2 行命令在后台运行 1 个 Alpine Linux 容器，通过执行 sleep 1d 命令保证容器可以持续运行。

```
4 $ docker run -it --rm --name app1 --hostname app1 --dns 202.102.224.68 --link db1:dbsrv1 alpine sh
```

第 4 行命令通过--hostname 选项设置容器的主机名，它会被写入容器内的/etc/hostname 和/etc/hosts 文件。通过--dns 选项设置 DNS 服务器，它会被写入容器内的/etc/resolv.conf 文件中，容器会用指定的 DNS 服务器来解析所有不在/etc/hosts 中的主机名。使用--link 选项添加指向另一个容器的链接，格式为 CONTAINER_NAME:ALIAS，在创建容器时，会将相关信息添加到容器内的/etc/hosts 文件中，所以新建容器可以使用别名、容器 ID 和容器名与另一个容器进行通信，示例命令如下：

```
/ # hostname
app1

/ # cat /etc/hostname
app1

/ # cat /etc/hosts
127.0.0.1 localhost
::1 localhost ip6-localhost ip6-loopback
fe00::0 ip6-localnet
ff00::0 ip6-mcastprefix
ff02::1 ip6-allnodes
ff02::2 ip6-allrouters
172.17.0.2 dbsrv1 29a5bbcbc24e db1
172.17.0.3 app1

/ # cat /etc/resolv.conf
nameserver 202.102.224.68
```

使用别名、容器名和容器 ID 进行连通性测试，示例命令如下：

```
/ # ping -c 2 dbsrv1
```

```
    PING dbsrv1 (172.17.0.2): 56 data bytes
    64 bytes from 172.17.0.2: seq=0 ttl=64 time=0.092 ms
    64 bytes from 172.17.0.2: seq=1 ttl=64 time=0.211 ms

    --- dbsrv1 ping statistics ---
    2 packets transmitted, 2 packets received, 0% packet loss
    round-trip min/avg/max = 0.092/0.151/0.211 ms

    / # ping -c 2 db1
    PING db1 (172.17.0.2): 56 data bytes
    64 bytes from 172.17.0.2: seq=0 ttl=64 time=0.066 ms
    64 bytes from 172.17.0.2: seq=1 ttl=64 time=0.204 ms

    --- db1 ping statistics ---
    2 packets transmitted, 2 packets received, 0% packet loss
    round-trip min/avg/max = 0.066/0.135/0.204 ms

    / # ping -c 2 29a5bbcbc24e
    PING 29a5bbcbc24e (172.17.0.2): 56 data bytes
    64 bytes from 172.17.0.2: seq=0 ttl=64 time=0.104 ms
    64 bytes from 172.17.0.2: seq=1 ttl=64 time=0.148 ms

    --- 29a5bbcbc24e ping statistics ---
    2 packets transmitted, 2 packets received, 0% packet loss
    round-trip min/avg/max = 0.104/0.126/0.148 ms
#使用 Ctrl+p、Ctrl+q 测试断开会话

5 $ docker ps -a
    CONTAINER ID   IMAGE    COMMAND      CREATED               STATUS               PORTS   NAMES
    38ec42806b44   alpine   "sh"         About a minute ago    Up About a minute            app1
    29a5bbcbc24e   alpine   "sleep 1d"   4 minutes ago         Up 4 minutes                 db1
```

10.3 容器的访问控制

10.3.1 容器访问外部网络

Docker 通过 iptables 防火墙软件来管理和控制容器网络。在图 10-2 中，容器 1 和容器 2 连接到默认的网桥 docker0 上，由于宿主机操作系统中启用了 IP 转发功能，又通过 iptables 实现了源 NAT，所以容器默认就可以访问外部网络。

下面通过实验对比启动 Docker 守护程序前后的变化，示例命令如下：

```
1 $ sudo systemctl disable docker

2 $ sudo reboot
```

设置 Docker 守护程序为手工启动，重新启动后查看系统状态，示例命令如下：

```
3 $ ip a
    1: lo: <LOOPBACK,UP,LOWER_UP> mtu 65536 qdisc noqueue state UNKNOWN group default qlen
      1000
        link/loopback 00:00:00:00:00:00 brd 00:00:00:00:00:00
        inet 127.0.0.1/8 scope host lo
            valid_lft forever preferred_lft forever
        inet6 ::1/128 scope host
            valid_lft forever preferred_lft forever
    2: ens160: <BROADCAST,MULTICAST,UP,LOWER_UP> mtu 1500 qdisc fq_codel state UP group
      default qlen 1000
        link/ether 00:0c:29:09:4d:e4 brd ff:ff:ff:ff:ff:ff
        altname enp3s0
        inet 192.168.1.231/24 brd 192.168.1.255 scope global noprefixroute ens160
            valid_lft forever preferred_lft forever
        inet6 fe80::20c:29ff:fe09:4de4/64 scope link noprefixroute
            valid_lft forever preferred_lft forever

4 $ nmcli connection
    NAME    UUID                                    TYPE      DEVICE
    ens160  3406a052-43de-3bd5-8d7c-6d57bf672e2e  ethernet  ens160

5 $ nmcli device
    DEVICE  TYPE      STATE      CONNECTION
    ens160  ethernet  connected  ens160
    lo      loopback  unmanaged  --
```

从第3~第5行命令的输出可以看出：在启动Docker守护程序之前，没有虚拟网桥docker0。

```
6 $ sysctl net.ipv4.ip_forward
    net.ipv4.ip_forward = 0

7 $ sudo iptables -t nat -L
    Chain PREROUTING (policy ACCEPT)
    target     prot opt source               destination

    Chain INPUT (policy ACCEPT)
    target     prot opt source               destination

    Chain OUTPUT (policy ACCEPT)
    target     prot opt source               destination

    Chain POSTROUTING (policy ACCEPT)
    target     prot opt source               destination
```

第6行命令的输出显示IP转发的状态，0表示没有开启转发功能。

第7行命令的输出显示宿主机的iptables中的NAT表是空的。

```
8 $ sudo systemctl start docker

9 $ sysctl net.ipv4.ip_forward
    net.ipv4.ip_forward = 1
```

Docker 守护程序命令 dockerd 的选项--ip-forward 默认值是 true，所以启动 Docker 守护程序时会自动配置宿主机的 IP 转发，这样第 9 行命令输出的状态值为 1。

```
10 $ ip a
    1: lo: <LOOPBACK,UP,LOWER_UP> mtu 65536 qdisc noqueue state UNKNOWN group default qlen 1000
        link/loopback 00:00:00:00:00:00 brd 00:00:00:00:00:00
        inet 127.0.0.1/8 scope host lo
            valid_lft forever preferred_lft forever
        inet6 ::1/128 scope host
            valid_lft forever preferred_lft forever
    2: ens160: <BROADCAST,MULTICAST,UP,LOWER_UP> mtu 1500 qdisc fq_codel state UP group
    default qlen 1000
        link/ether 00:0c:29:09:4d:e4 brd ff:ff:ff:ff:ff:ff
        altname enp3s0
        inet 192.168.1.231/24 brd 192.168.1.255 scope global noprefixroute ens160
            valid_lft forever preferred_lft forever
        inet6 fe80::20c:29ff:fe09:4de4/64 scope link noprefixroute
            valid_lft forever preferred_lft forever
    3: docker0: <NO-CARRIER,BROADCAST,MULTICAST,UP> mtu 1500 qdisc noqueue state DOWN
    group default
        link/ether 02:42:8b:92:bc:ff brd ff:ff:ff:ff:ff:ff
        inet 172.17.0.1/16 brd 172.17.255.255 scope global docker0
            valid_lft forever preferred_lft forever

11 $ nmcli connection
    NAME       UUID                                    TYPE       DEVICE
    ens160     3406a052-43de-3bd5-8d7c-6d57bf672e2ee   thernet    ens160
    docker0    d3af44ee-c29a-4b09-9e31-3792acf42670    bridge     docker0

12 $ nmcli device
    DEVICE     TYPE       STATE                    CONNECTION
    ens160     ethernet   connected                ens160
    docker0    bridge     connected (externally)   docker0
    lo         loopback   unmanaged                --
```

第 10~第 12 行命令的输出显示：宿主机中新增一个名为 docker0 的虚拟网桥，它的 IP 地址是 172.17.0.1/16。

```
13 $ sudo iptables -t nat -L -n
    Chain PREROUTING (policy ACCEPT)
    target     prot opt source          destination
    DOCKER     all  --  0.0.0.0/0       0.0.0.0/0          ADDRTYPE match dst-type LOCAL

    Chain INPUT (policy ACCEPT)
    target     prot opt source          destination

    Chain OUTPUT (policy ACCEPT)
    target     prot opt source          destination
    DOCKER     all  --  0.0.0.0/0       !127.0.0.0/8       ADDRTYPE match dst-type LOCAL

    Chain POSTROUTING (policy ACCEPT)
```

```
target          prot opt  source          destination
MASQUERADE      all  --   172.17.0.0/16   0.0.0.0/0

Chain DOCKER (2 references)
target     prot opt  source       destination
RETURN     all  --   0.0.0.0/0    0.0.0.0/0
```

第 13 行命令的输出显示 iptables 中的 NAT 的规则。从规则可以看出：对容器的网络 172.17.0.0/16 将通过 MASQUERADE 实现源 NAT。

还可以在系统日志中看到与启动 Docker 守护程序有关的事件，示例信息如下：

```
systemd[1]: Starting Docker Socket for the API...
systemd[1]: Starting containerd container runtime...
systemd[1]: Listening on Docker Socket for the API.
containerd[1407]: time="2022-12-10T10:40:05.785880257+08:00" level=info msg="starting containerd"
  revision=1c90a442489720eec95342e1789ee8a5e1b9536f version=1.6.9
containerd[1407]: time="2022-12-10T10:40:05.798228964+08:00" level=info msg="loading plugin
  \"io.containerd.content.v1.content\"..." type=io.containerd.content.v1

systemd[1]: Started containerd container runtime.
systemd[1]: Starting Docker Application Container Engine...

kernel: Bridge firewalling registered
dockerd[1413]: time="2022-12-10T10:40:06.066848839+08:00" level=info msg="Firewalld: docker zone already
  exists, returning"
kernel: Warning: Deprecated Driver is detected: nft_compat will not be maintained in a future major release and
  may be disabled
firewalld[753]: WARNING: COMMAND_FAILED: '/usr/sbin/iptables -w10 -t nat -D PREROUTING -m
  addrtype --dst-type LOCAL -j DOCKER' failed: iptables v1.8.8 (nf_tables): Chain 'DOCKER' does not
  exist#012Try `iptables -h' or 'iptables --help' for more information.
firewalld[753]: WARNING: COMMAND_FAILED: '/usr/sbin/iptables -w10 -t nat -D OUTPUT -m addrtype
  --dst-type LOCAL ! --dst 127.0.0.0/8 -j DOCKER' failed: iptables v1.8.8 (nf_tables): Chain 'DOCKER' does not
  exist#012Try `iptables -h' or 'iptables --help' for more information.

dockerd[1413]: time="2022-12-10T10:40:06.578199549+08:00" level=info msg="Default bridge (docker0) is
  assigned with an IP address 172.17.0.0/16. Daemon option --bip can be used to set a preferred IP address"

dockerd[1413]: time="2022-12-10T10:40:06.789715163+08:00" level=info msg="Daemon has completed
  initialization"
systemd[1]: Started Docker Application Container Engine.
```

创建一个新容器，在容器中检查是否可以访问外部网络，示例命令如下：

```
$ docker run --rm -it alpine sh
  / # ip a
  1: lo: <LOOPBACK,UP,LOWER_UP> mtu 65536 qdisc noqueue state UNKNOWN qlen 1000
       link/loopback 00:00:00:00:00:00 brd 00:00:00:00:00:00
       inet 127.0.0.1/8 scope host lo
          valid_lft forever preferred_lft forever
  4: eth0@if5: <BROADCAST,MULTICAST,UP,LOWER_UP,M-DOWN> mtu 1500 qdisc noqueue state UP
       link/ether 02:42:ac:11:00:02 brd ff:ff:ff:ff:ff:ff
```

```
    inet 172.17.0.2/16 brd 172.17.255.255 scope global eth0
       valid_lft forever preferred_lft forever

/ # ip r
default via 172.17.0.1 dev eth0
172.17.0.0/16 dev eth0 scope link   src 172.17.0.2

/ # ping -c 2 8.8.8.8
PING 8.8.8.8 (8.8.8.8): 56 data bytes
64 bytes from 8.8.8.8: seq=0 ttl=113 time=39.036 ms
64 bytes from 8.8.8.8: seq=1 ttl=113 time=38.925 ms

--- 8.8.8.8 ping statistics ---
2 packets transmitted, 2 packets received, 0% packet loss
round-trip min/avg/max = 38.925/38.980/39.036 ms
```

容器默认网关为 docker0 网桥上的接口 172.17.0.1，由于宿主机提供了源 NAT 功能，所以容器可以访问外部网络(如 8.8.8.8)。

10.3.2 容器之间相互访问

Docker 守护程序的命令 dockerd 有一个名为 icc 的选项，它是容器内通信(inter container communication)的缩写，默认值是 true。由于容器默认连接到 docker0 网桥上，所以它们之间是可以相互访问的。

可以通过 icc 选项来控制容器之间的相互访问，示例命令如下：

```
1 $ sudo iptables -L -n
    ……略……
    Chain FORWARD (policy DROP)
    target                      prot opt source        destination
    DOCKER-USER                 all  --  0.0.0.0/0     0.0.0.0/0
    DOCKER-ISOLATION-STAGE-1    all  --  0.0.0.0/0     0.0.0.0/0
    ACCEPT                      all  --  0.0.0.0/0     0.0.0.0/0    ctstate RELATED,ESTABLISHED
    DOCKER                      all  --  0.0.0.0/0     0.0.0.0/0
    ACCEPT                      all  --  0.0.0.0/0     0.0.0.0/0
    ACCEPT                      all  --  0.0.0.0/0     0.0.0.0/0
    ……
```

使用第 1 行命令查看当前 iptables 的 filter 表设置。如果 icc 值是 true，则会在 FORWARD 链上增加 ACCEPT 策略，从而允许容器之间的通信。

```
2 $ sudo vi /etc/docker/daemon.json
    {
        "icc" : false
    }

3 $ sudo systemctl restart docker


4 $ sudo iptables -L -n
```

```
……
Chain FORWARD (policy DROP)
target                          prot opt source        destination
DOCKER-USER                     all  --  0.0.0.0/0     0.0.0.0/0
DOCKER-ISOLATION-STAGE-1all  --  0.0.0.0/0     0.0.0.0/0
ACCEPT                          all  --  0.0.0.0/0     0.0.0.0/0 ctstate RELATED,ESTABLISHED
DOCKER                          all  --  0.0.0.0/0     0.0.0.0/0
ACCEPT                          all  --  0.0.0.0/0     0.0.0.0/0
DROP                            all  --  0.0.0.0/0     0.0.0.0/0
……略……
```

第 2 行命令修改 Docker 守护程序的配置文件，将 icc 选项设置为 false。

重新启动 Docker 守护程序之后，通过第 4 行命令再次查看 FORWARD 链中的规则，原来的 ACCEPT 策略被替换为 DROP 策略。

```
5 $ docker run --rm --name c1 -itd alpine sh

6 $ docker inspect c1 | grep 172.17
                    "Gateway": "172.17.0.1",
                    "IPAddress": "172.17.0.2",
                            "Gateway": "172.17.0.1",
                            "IPAddress": "172.17.0.2",

7 $ ping -c 2 172.17.0.2
    PING 172.17.0.2 (172.17.0.2) 56(84) bytes of data.
    64 bytes from 172.17.0.2: icmp_seq=1 ttl=64 time=0.046 ms
    64 bytes from 172.17.0.2: icmp_seq=2 ttl=64 time=0.129 ms

    --- 172.17.0.2 ping statistics ---
    2 packets transmitted, 2 received, 0% packet loss, time 1029ms
    rtt min/avg/max/mdev = 0.046/0.087/0.129/0.041 ms

8 $ docker run --rm --name c2 -it alpine ping -c 2 172.17.0.2
    PING 172.17.0.2 (172.17.0.2): 56 data bytes

    --- 172.17.0.2 ping statistics ---
    2 packets transmitted, 0 packets received, 100% packet loss
```

第 5 行命令运行一个名为 c1 的容器。通过第 6 行命令获得它的 IP 地址。

从第 7 行命令输出可以看出：宿主机可以与容器 c1(172.17.0.2)进行通信。

第 8 行命令再创建新的容器 c2。从 ping 命令的输出可以看出：容器 c2 无法与容器 c1 进行通信，从而达到类似私有 VLAN(private VLAN)的效果，容器可以访问外部网络，但是容器之间不能相互访问。

从上面的实验可以看出：通过 icc 选项要么允许容器之间进行通信，要么禁止它们之间进行通信。是否可以进行细粒度的灵活配置呢？可以且有多种方法。其中一种方法是设置 FORWARD 链的规则，例如：仅允许 172.17.0.2 与 172.17.0.3 之间进行通信，示例命令如下：

```
1 $ sudo iptables -L FORWARD --line-number -n
    Chain FORWARD (policy DROP)
```

```
    num  target                       prot opt source     destination
    1    DOCKER-USER                  all  --  0.0.0.0/0  0.0.0.0/0
    2    DOCKER-ISOLATION-STAGE-1 all  --  0.0.0.0/0  0.0.0.0/0
    3    ACCEPT                       all  --  0.0.0.0/0  0.0.0.0/0  ctstate RELATED,ESTABLISHED
    4    DOCKER                       all  --  0.0.0.0/0  0.0.0.0/0
    5    ACCEPT                       all  --  0.0.0.0/0  0.0.0.0/0
    6    DROP                         all  --  0.0.0.0/0  0.0.0.0/0

2 $ sudo iptables -I FORWARD 1 -s 172.17.0.3 -d 172.17.0.2 -j ACCEPT

3 $ sudo iptables -I FORWARD 1 -s 172.17.0.2 -d 172.17.0.3 -j ACCEPT

4 $ sudo iptables -L FORWARD --line-number -n
    Chain FORWARD (policy DROP)
    num  target          prot opt source         destination
    1    ACCEPT          all  --  172.17.0.2     172.17.0.3
    2    ACCEPT          all  --  172.17.0.3     172.17.0.2
    3    DOCKER-USER     all  --  0.0.0.0/0      0.0.0.0/0
    4    DOCKER-ISOLATION-STAGE-1    all  --  0.0.0.0/0        0.0.0.0/0
    5    ACCEPT          all  --  0.0.0.0/0    0.0.0.0/0      ctstate RELATED,ESTABLISHED
    6    DOCKER          all  --  0.0.0.0/0    0.0.0.0/0
    7    ACCEPT          all  --  0.0.0.0/0    0.0.0.0/0
    8    DROP            all  --  0.0.0.0/0    0.0.0.0/0

5 $ docker run --rm --name c2 -it alpine sh

    / # ip a
    1: lo: <LOOPBACK,UP,LOWER_UP> mtu 65536 qdisc noqueue state UNKNOWN qlen 1000
        link/loopback 00:00:00:00:00:00 brd 00:00:00:00:00:00
        inet 127.0.0.1/8 scope host lo
           valid_lft forever preferred_lft forever
    8: eth0@if9: <BROADCAST,MULTICAST,UP,LOWER_UP,M-DOWN> mtu 1500 qdisc noqueue state UP
        link/ether 02:42:ac:11:00:03 brd ff:ff:ff:ff:ff:ff
        inet 172.17.0.3/16 brd 172.17.255.255 scope global eth0
           valid_lft forever preferred_lft forever

    / # ping -c 2 172.17.0.2
    PING 172.17.0.2 (172.17.0.2): 56 data bytes
    64 bytes from 172.17.0.2: seq=0 ttl=64 time=0.094 ms
    64 bytes from 172.17.0.2: seq=1 ttl=64 time=0.271 ms

    --- 172.17.0.2 ping statistics ---
    2 packets transmitted, 2 packets received, 0% packet loss
    round-trip min/avg/max = 0.094/0.182/0.271 ms
    # 按下 Ctrl+p、Ctrl+q 组合键盘出
```

容器 c1 的地址是 172.17.0.2，容器 c2 的地址是 172.17.0.3，所以可以同时访问。

```
7 $ docker run --rm --name c3 -it alpine sh

    / # ip a
```

```
1: lo: <LOOPBACK,UP,LOWER_UP> mtu 65536 qdisc noqueue state UNKNOWN qlen 1000
    link/loopback 00:00:00:00:00:00 brd 00:00:00:00:00:00
    inet 127.0.0.1/8 scope host lo
       valid_lft forever preferred_lft forever
10: eth0@if11: <BROADCAST,MULTICAST,UP,LOWER_UP,M-DOWN> mtu 1500 qdisc noqueue state UP
    link/ether 02:42:ac:11:00:04 brd ff:ff:ff:ff:ff:ff
    inet 172.17.0.4/16 brd 172.17.255.255 scope global eth0
       valid_lft forever preferred_lft forever

/ # ping -c 2 172.17.0.2
PING 172.17.0.2 (172.17.0.2): 56 data bytes

--- 172.17.0.2 ping statistics ---
2 packets transmitted, 0 packets received, 100% packet loss
```

容器 c3 的地址是 172.17.0.4，iptables 防火墙不允许访问容器 c1。这种方法实现了对容器之间细粒度的访问控制。

10.4 容器的端口映射

在默认情况下，容器可以主动访问宿主机外面的网络。如果允许宿主机外面的计算机能够访问容器中的资源，最便捷的方法就是使用端口映射。端口映射也被称为“服务对外发布”。

使用 create 或 run 子命令时，可以通过两个选项来映射端口：

- -p, --publish：发布指定的端口。
- -P, --publish-all：将所有公开的端口发布到随机端口。

对外发布 HTTPD 的端口，示例命令如下：

```
1 $ docker image inspect httpd
   ......
           "ExposedPorts": {
               "80/tcp": {}
           },
   ......

2 $ docker run --rm -d --name web1 -P httpd

3 $ docker ps
    CONTAINER ID    IMAGE    COMMAND    CREATED    STATUS
    ff664295c100    httpd    "httpd-foreground"    6 seconds ago    Up 4 seconds
PORTS                                          NAMES
0.0.0.0:49153->80/tcp, :::49153->80/tcp    web1


4 $ docker port web1
    80/tcp -> 0.0.0.0:49153
    80/tcp -> :::49153
```

```
5 $ curl http://localhost:49153
    <html><body><h1>It works!</h1></body></html>
```

从第 1 行命令的输出中可以看出：镜像 HTTPD 对外暴露的端口是 TCP 80。

第 2 行命令使用-P 选项运行容器，Docker 守护程序会自动在本地 49000～49900 范围内随机选择 1 个未被占用的端口，将它映射到容器的 80 端口。

提示

不同的 Docker 版本会有一定的差异，例如：在 v20 版本中，第 1 个映射端口为 49153，第 2 个是 49154，以此类推。

从第 3、第 4 行命令输出中可看出：宿主机的 49153 端口被映射到容器的 80 端口。

通过第 5 行命令测试对 localhost:49153 的访问。在外部网络，通过浏览器可以正常访问容器中的 Web 服务，如图 10-3 所示。

图 10-3　访问容器中的 Web 服务

容器的端口映射通过在宿主机 iptables 的 NAT 表中添加 DNAT 策略来实现。查看宿主机 iptables 中的规则，示例命令如下：

```
6 $ sudo iptables -t nat -nvL
    ……
    Chain POSTROUTING (policy ACCEPT 0 packets, 0 bytes)
     pkts bytes target     prot opt in   out      source          destination
        0     0 MASQUERADE  all  --  *  !docker0  172.17.0.0/16  0.0.0.0/0
        0     0 MASQUERADE  tcp  --  *   *        172.17.0.2     172.17.0.2     tcp dpt:80

    Chain DOCKER (2 references)
     pkts bytes target     prot opt in        out   source      destination
        0     0 RETURN     all  --  docker0   *    0.0.0.0/0   0.0.0.0/0
        2   104 DNAT       tcp  --  !docker0  *    0.0.0.0/0   0.0.0.0/0   tcp dpt:49153 to:172.17.0.2:80

7 $ sudo iptables -L -n
    ……
    Chain DOCKER (1 references)
    target     prot opt source        destination
    ACCEPT     tcp  --  0.0.0.0/0     172.17.0.2          tcp dpt:80
```

除了将所有公开的端口发布到随机端口，还可以用-p 选项发布指定端口，格式为：

ip:[hostPort]:containerPort | [hostPort:]containerPort。如果希望将宿主机的 80 映射为容器的 80 端口，示例命令如下：

```
8 $ docker run --rm -d --name web2 -p 80:80 nginx

9 $ docker ps -a
    CONTAINER ID    IMAGE    COMMAND    CREATED    STATUS
    2d34d0421bd6    nginx    "/docker-entrypoint.…"    4 seconds ago    Up 3 seconds
     PORTS    NAMES
    ff664295c100    httpd    "httpd-foreground"    19 minutes ago    Up 19 minutes
0.0.0.0:80->80/tcp, :::80->80/tcp    web2
0.0.0.0:49153->80/tcp, :::49153->80/tcp    web1
```

第 9 行命令输出显示：映射地址为 0.0.0.0，这意味着将接受来自所有网络接口上的流量。如果希望进行细粒度控制，可以通过-p IP:host_port:container_port 或-p IP::port 来指定特定外部网络接口地址。如果希望自动映射到某个固定的宿主机 IP 地址，可以在 Docker 配置文件中指定 DOCKER_OPTS="--ip=IP_ADDRESS"，然后重启 Docker 服务以便生效。

10.5 容器的便捷互联机制

在使用 create 或 run 子命令时，可以使用--link 选项实现便捷互联机制来简化管理，它的主要功能有：

- 在/etc/hosts 中添加容器名称、容器 ID 和别名的映射。
- 获得另外容器的环境变量。

下面运行两个容器进行简单的测试，示例命令如下：

```
1 $ docker pull mysql

2 $ docker image inspect mysql | more
    ……
    "Env": [
        "PATH=/usr/local/sbin:/usr/local/bin:/usr/sbin:/usr/bin:/sbin:/bin",
        "GOSU_VERSION=1.14",
        "MYSQL_MAJOR=8.0",
        "MYSQL_VERSION=8.0.29-1.el8",
        "MYSQL_SHELL_VERSION=8.0.29-1.el8"
    ],
    ……
```

从第 2 行命令的输出可以看出：在 mysql 镜像中定义了多个环境变量。

```
3 $ docker run -d --rm --name db1 mysql sleep 1d

4 $ docker ps -a
    CONTAINER ID  IMAGE    COMMAND    CREATED    STATUS    PORTS    NAMES
    b636b24d0ecd    mysql  "docker-entrypoint.s…"  4 seconds ago  Up 3 seconds 3306/tcp, 33060/tcp db1
```

```
5 $ docker exec db1 env
    PATH=/usr/local/sbin:/usr/local/bin:/usr/sbin:/usr/bin:/sbin:/bin
    HOSTNAME=b636b24d0ecd
    GOSU_VERSION=1.14
    MYSQL_MAJOR=8.0
    MYSQL_VERSION=8.0.29-1.el8
    MYSQL_SHELL_VERSION=8.0.29-1.el8
    HOME=/root
```

从第 5 行命令的输出可以看出：第 1 个容器 db1 中有多个环境变量。

使用--link 选项运行第 2 个容器，示例命令如下：

```
6 $ docker run -it –rm --name web1 --link db1:dbsrv1 alpine sh

    / # cat /etc/hosts
    127.0.0.1           localhost
    ……
    172.17.0.2          dbsrv1 b636b24d0ecd db1
    172.17.0.3          fb756a537b2d

    / # env
    HOSTNAME=fb756a537b2d
    SHLVL=1
    HOME=/root
    PATH=/usr/local/sbin:/usr/local/bin:/usr/sbin:/usr/bin:/sbin:/bin
    PWD=/
    DBSRV1_ENV_MYSQL_MAJOR=8.0
    DBSRV1_PORT_33060_TCP=tcp://172.17.0.2:33060
    DBSRV1_PORT_3306_TCP_ADDR=172.17.0.2
    DBSRV1_ENV_GOSU_VERSION=1.14
    DBSRV1_PORT_3306_TCP_PORT=3306
    DBSRV1_PORT_3306_TCP_PROTO=tcp
    DBSRV1_PORT_33060_TCP_ADDR=172.17.0.2
    DBSRV1_PORT=tcp://172.17.0.2:3306
    DBSRV1_PORT_3306_TCP=tcp://172.17.0.2:3306
    DBSRV1_PORT_33060_TCP_PORT=33060
    DBSRV1_ENV_MYSQL_VERSION=8.0.29-1.el8
    DBSRV1_PORT_33060_TCP_PROTO=tcp
    DBSRV1_NAME=/web1/dbsrv1
    DBSRV1_ENV_MYSQL_SHELL_VERSION=8.0.29-1.el8

    / # ping -c 2   db1
    PING db1 (172.17.0.2): 56 data bytes
    64 bytes from 172.17.0.2: seq=0 ttl=64 time=0.526 ms
    64 bytes from 172.17.0.2: seq=1 ttl=64 time=0.128 ms

    --- db1 ping statistics ---
    2 packets transmitted, 2 packets received, 0% packet loss
    round-trip min/avg/max = 0.128/0.327/0.526 m
```

第 6 行命令中指定--link 选项的格式为 name-or-id[:alias]，给容器 db1 增加一个名为 dbsrv1 的别名。在第 2 个容器的/etc/hosts 文件中有容器名称、容器 ID 和别名的记录，同时也有多个与容器 db1 相关的环境变量。

说明

Link 标志是 Docker 的传统功能，将来的版本中可能会移除这个功能。建议通过自定义网络来简化容器之间的通信，通过卷的机制在容器之间共享信息。

10.6 容器的网络管理命令

如果有更复杂的网络需求，就需要使用网络管理命令，docker network 主要包括以下子命令：

- create：创建自定义网络。
- connect：将容器接入到网络。
- disconnect：将容器从网络上断开。
- inspect：查看网络的详细信息。
- ls：列出所有的网络。
- prune：清理无用的网络。
- rm：删除网络。

10.6.1 列出网络

使用 list 子命令列出 Docker 守护程序知道的网络，它的语法格式如下：

```
docker network ls [OPTIONS]
```

选项：

- -f, --filter：指定输出过滤器，如 driver=bridge。
- --format：通过 Go 语言格式的模板控制输出的内容。
- --no-trunc：不截断输出内容。
- -q, --quiet：安静模式，只输出网络的 ID。

查看宿主机当前的容器网络，示例命令如下：

```
1 $ docker network ls
    NETWORK ID      NAME        DRIVER      SCOPE
    6e03f8c40591    bridge      bridge      local
    3b86c5783bd6    host        host        local
    7aaea75e88a4    none        null        local

2 $ docker network ls --filter driver=bridge --no-trunc
    NETWORK ID                                                          NAME DRIVER  SCOPE
    6e03f8c4059195883b19648d2c13c02066b836699f672ef63907786634e0a608   bridge  bridge  local
```

第 1 行命令列出所有网络。有 3 种驱动类型的网络：bridge、host 和 null，其中 bridge 是常用的网络类型，host 用于容器直接连接宿主机外部网络，none 用于没有网络的容器。

第 2 行命令仅列出使用 bridge 驱动程序的网络，通过--no-trunc 选项显示完整的网络 ID。

10.6.2 查看网络信息

使用 inspect 命令可以查看网络的具体信息(JSON 格式)，包括接入的容器、网络配置信息等，它的语法格式如下：

```
docker network inspect [OPTIONS] NETWORK [NETWORK...]
```

选项：

- -f, --format：通过 Go 语言格式的模板控制输出内容。
- -v, --verbose：输出详细信息用于诊断。

查看 bridge 网络的信息，示例命令如下：

```
$ docker network inspect bridge
    [
        {
            "Name": "bridge",
......
            "IPAM": {
                "Driver": "default",
                "Options": null,
                "Config": [
                    {
                        "Subnet": "172.17.0.0/16",
                        "Gateway": "172.17.0.1"
                    }
                ]
            },
......
$ docker network inspect --format  '{{.IPAM.Config}}' bridge
[{172.17.0.0/16  172.17.0.1 map[]}]
```

下面启动两个 Alpine Linux 容器，考察默认的 bridge 网络及容器之间的通信，示例命令如下：

```
1 $ docker run --rm -itd --name a1 alpine sh

2 $ docker run --rm -itd --name a2 alpine sh

3 $ docker ps -a
    CONTAINER ID  IMAGE   COMMAND  CREATED         STATUS        PORTS    NAMES
    296a491c433c  alpine  "sh"     3 seconds ago   Up 3 seconds  a2
    5306d4549d9e  alpine  "sh"     9 seconds ago   Up 9 seconds  a1
```

运行容器时没有使用--network 选项，所以两个容器都会连接到默认 bridge 网络。考察宿主机网络的变化，示例命令如下：

```
4 $ ip addr
    ......略......
    3: docker0: <BROADCAST,MULTICAST,UP,LOWER_UP> mtu 1500 qdisc noqueue state UP group default
        link/ether 02:42:c8:03:fc:9a brd ff:ff:ff:ff:ff:ff
```

```
    inet 172.17.0.1/16 brd 172.17.255.255 scope global docker0
       valid_lft forever preferred_lft forever
    inet6 fe80::42:c8ff:fe03:fc9a/64 scope link
       valid_lft forever preferred_lft forever
11: vethb96b21c@if10: <BROADCAST,MULTICAST,UP,LOWER_UP> mtu 1500 qdisc noqueue master
  docker0 state UP group default
    link/ether 76:c8:79:3d:d3:47 brd ff:ff:ff:ff:ff:ff link-netnsid 0
    inet6 fe80::74c8:79ff:fe3d:d347/64 scope link
       valid_lft forever preferred_lft forever
13: veth956f6ec@if12: <BROADCAST,MULTICAST,UP,LOWER_UP> mtu 1500 qdisc noqueue master
  docker0 state UP group default
    link/ether 6e:41:ad:03:67:43 brd ff:ff:ff:ff:ff:ff link-netnsid 1
    inet6 fe80::6c41:adff:fe03:6743/64 scope link
       valid_lft forever preferred_lft forever

5 $ nmcli device
DEVICE        TYPE       STATE                    CONNECTION
ens160        ethernet   connected                ens160
docker0       bridge     connected (externally)   docker0
veth956f6ec   ethernet   unmanaged                --
vethb96b21c   ethernet   unmanaged                --
lo            loopback   unmanaged                --
```

考察 bridge 网络，示例命令如下：

```
6 $ docker network inspect bridge
    ……
    "Containers": {
        "296a491c433c75bfb78f6954bb5043db7827e3498a5630a618b42826c5a94b71": {
            "Name": "a2",
            "EndpointID": "e8d4ee51bc33a0cabdbdff665a6511b5d9bcab6c9209f8b6910c8aefa50c2117",
            "MacAddress": "02:42:ac:11:00:03",
            "IPv4Address": "172.17.0.3/16",
            "IPv6Address": ""
        },
        "5306d4549d9e62f0b49694678298414e92654d66db07e7877da2f710c0082061": {
            "Name": "a1",
            "EndpointID": "9c0953172467f0880330bdc1eed300624810b2a33325f5e4b2fe18710e6c99a6",
            "MacAddress": "02:42:ac:11:00:02",
            "IPv4Address": "172.17.0.2/16",
            "IPv6Address": ""
        }
    },
    ……
```

第 6 行命令的输出显示：在 Containers 键会包括每个容器的信息，包括 MAC 地址和 IP 地址的信息，其中容器 a1 为 172.17.0.2、容器 a2 为 172.17.0.3。

连接到容器 a1 的控制台，考察容器之间的通信，示例命令如下：

```
7 $ docker attach a1
    / # cat /etc/hosts
```

```
127.0.0.1 localhost
::1 localhost ip6-localhost ip6-loopback
fe00::0 ip6-localnet
ff00::0 ip6-mcastprefix
ff02::1 ip6-allnodes
ff02::2 ip6-allrouters
172.17.0.2 5306d4549d9e

/ # ping -c 2 172.17.0.3
PING 172.17.0.3 (172.17.0.3): 56 data bytes
64 bytes from 172.17.0.3: seq=0 ttl=64 time=0.169 ms
64 bytes from 172.17.0.3: seq=1 ttl=64 time=0.127 ms

--- 172.17.0.3 ping statistics ---
2 packets transmitted, 2 packets received, 0% packet loss
round-trip min/avg/max = 0.127/0.148/0.169 ms

/ # ping -c 2 a2
ping: bad address 'a2'
```

从上述实验可以看出：在默认的 bridge 网络中，无法使用容器的名称来进行网络访问。除了使用--link 选项在/etc/hosts 中添加新的映射之外，还可以通过自定义容器网络来实现。

10.6.3 创建自定义网络

在正式的生产环境中，建议使用自定义网络而非默认的 bridge 网络，如图 10-4 所示。默认情况下，自定义网络中的容器可以与外部网络进行通信，可以与同一网络中的容器进行通信，但是不同网络之间容器不能进行通信。同时 Docker 还为自定义网络提供 DNS 服务，容器之间可以使用容器名称、容器 ID 和主机名进行互相访问，这样既不需要添加/etc/hosts 中的映射，也不需要使用--link 选项。

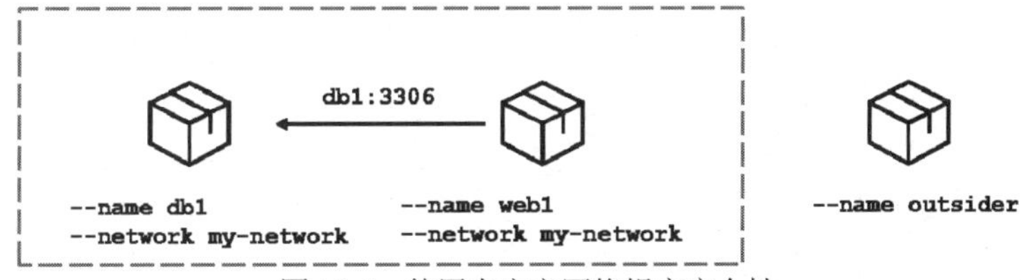

图 10-4　使用自定义网络提高安全性

使用 create 子命令可以创建自定义的网络，它的语法格式如下：

```
docker network create [OPTIONS] NETWORK
```

选项：

- --attachable：支持手动容器挂载。
- --aux-address：辅助的 IP 地址。
- --config-from：从指定的网络复制配置数据。

- --config-only：启用仅可配置模式。
- -d,--driver：网络驱动类型，默认为 bridge。
- --gateway：网关地址。
- --ingress：创建一个 Swarm 可路由的网状网络。
- --internal：内部模式，禁止外部对所创建网络的访问。
- --ip-range：指定分配 IP 地址的范围。
- --ipam-driver：IP 地址管理的插件类型。
- --ipam-opt：IP 地址管理插件的选项。
- --ipv6：支持 IPv6 地址。
- --label：为网络添加元标签信息。
- -o,--opt：网络驱动所支持的选项。
- --scope：指定网络范围。
- --subnet：网络地址段，CIDR 格式，如 172.17.0.0/16。

创建使用 bridge 驱动的名为 my-network 的自定义网络，示例命令如下：

```
1 $ docker network create --driver bridge my-network

2 $ docker network ls
    NETWORK ID        NAME           DRIVER    SCOPE
    1a95ed5f503f      bridge         bridge    local
    3b86c5783bd6      host           host      local
    dd67501d1070      my-network     bridge    local
    7aaea75e88a4      none           null      local

3 $ docker network inspect my-network
    [
        {
            "Name": "my-network",
            "Id": "dd67501d1070ddc93c26c3dbf9378f845ffb602a6a2d415eb6c39aa0a093c8cd",
            "Created": "2022-12-12T15:31:09.171655731+08:00",
            "Scope": "local",
            "Driver": "bridge",
            "EnableIPv6": false,
            "IPAM": {
                "Driver": "default",
                "Options": {},
                "Config": [
                    {
                        "Subnet": "172.18.0.0/16",
                        "Gateway": "172.18.0.1"
                    }
                ]
            },
            "Internal": false,
            "Attachable": false,
            "Ingress": false,
            "ConfigFrom": {
```

```
                "Network": ""
            },
            "ConfigOnly": false,
            "Containers": {},
            "Options": {},
            "Labels": {}
        }
    ]
```

从第 3 行命令的输出可以看出：新网络的 IP 地址段是 172.18.0.0/16。

查看宿主机网络的变化，示例命令如下：

```
4 $ nmcli device
    DEVICE              TYPE        STATE                     CONNECTION
    ens160              ethernet    connected                 ens160
    br-dd67501d1070     bridge      connected (externally)    br-dd67501d1070
    docker0             bridge      connected (externally)    docker0
    lo                  loopback    unmanaged                 --

5 $ ip a
    ......
    3: docker0: <NO-CARRIER,BROADCAST,MULTICAST,UP> mtu 1500 qdisc noqueue state DOWN
      group default
        link/ether 02:42:45:ff:91:8f brd ff:ff:ff:ff:ff:ff
        inet 172.17.0.1/16 brd 172.17.255.255 scope global docker0
           valid_lft forever preferred_lft forever
    4: br-dd67501d1070: <NO-CARRIER,BROADCAST,MULTICAST,UP> mtu 1500 qdisc noqueue state
      DOWN group default
        link/ether 02:42:fb:0b:8e:76 brd ff:ff:ff:ff:ff:ff
        inet 172.18.0.1/16 brd 172.18.255.255 scope global br-dd67501d1070
           valid_lft forever preferred_lft forever
```

查看宿主机中 iptables 规则的变化，示例命令如下：

```
6 $ sudo iptables -nvL -t nat
    ......
    Chain POSTROUTING (policy ACCEPT 0 packets, 0 bytes)
     pkts  bytes  target        prot  opt  in   out                source          destination
        0      0  MASQUERADE    all   --   *    !br-dd67501d1070   172.18.0.0/16   0.0.0.0/0
        0      0  MASQUERADE    all   --   *    !docker0           172.17.0.0/16   0.0.0.0/0

    Chain DOCKER (2 references)
     pkts  bytes  target    prot opt  in   out                      source       destination
        0  0      RETURN    all  --   br-dd67501d1070 *             0.0.0.0/0    0.0.0.0/0
        0  0      RETURN    all  --   docker0 *                     0.0.0.0/0    0.0.0.0/0
```

从第 6 行命令的输出可以看出：新增针对网络 172.18.0.0/16 的 SNAT(MASQUERADE)规则，所以这个网络中的容器可以访问宿主机的外部网络。

使用--network 选项创建两个 my-network 的容器，示例命令如下：

```
7 $ docker run --rm -itd --name db1 --network my-network alpine sh
```

```
8 $ docker run --rm -itd --name web1 --network my-network alpine sh

9 $ docker network inspect my-network
    ……
    "Containers": {
        "04af941dc86315d7405617ccc5d60195be428af1a51830fcea4515b26c6984d5": {
            "Name": "web1",
            "EndpointID": "5dc358c90c684aa6dcc52089263a1bed1ebb1ebcbcbfbead3b807aaeea158fe7",
            "MacAddress": "02:42:ac:12:00:03",
            "IPv4Address": "172.18.0.3/16",
            "IPv6Address": ""
        },
        "55e961ed457706bb438a9b24cb578bb7eed9adb817ed7361905ab00ab9cde7ed": {
            "Name": "db1",
            "EndpointID": "85572965960d1e261b5ba7d4ff39237f11e5f131ff903448d4ccac8e0052858f",
            "MacAddress": "02:42:ac:12:00:02",
            "IPv4Address": "172.18.0.2/16",
            "IPv6Address": ""
        }
    },
    ……
```

从第 9 行命令的输出可以看出：容器 db1 的 IP 地址是 172.18.0.2/16，容器 web1 的 IP 地址是 172.18.0.3/16。

附加到容器 web1 的控制台，验证网络的连通性，示例命令如下：

```
10 $ docker exec -it web1 sh

    / # ip r
    default via 172.18.0.1 dev eth0
    172.18.0.0/16 dev eth0 scope link    src 172.18.0.3

    / # ping -c 2 www.baidu.com
    PING www.baidu.com (220.181.38.149): 56 data bytes
    64 bytes from 220.181.38.149: seq=0 ttl=127 time=24.393 ms
    64 bytes from 220.181.38.149: seq=1 ttl=127 time=24.604 ms

    --- www.baidu.com ping statistics ---
    2 packets transmitted, 2 packets received, 0% packet loss
    round-trip min/avg/max = 24.393/24.498/24.604 ms

    / # cat /etc/hosts
    127.0.0.1 localhost
    ::1 localhost ip6-localhost ip6-loopback
    fe00::0 ip6-localnet
    ff00::0 ip6-mcastprefix
    ff02::1 ip6-allnodes
    ff02::2 ip6-allrouters
    172.18.0.3 04af941dc863

    / # cat /etc/resolv.conf
```

```
nameserver 127.0.0.11
options ndots:0
```

自定义网络中的容器不再使用宿主机的 DNS 服务器配置，而是使用 Docker 内置的 DNS 服务器(127.0.0.11)，它既可以解析外部网络的域名，也可以解析自定义网络中的容器。

使用容器的 ID、容器的名称以及主机名进行连通性测试，示例命令如下：

```
/ # ping -c 2 db1
    PING db1 (172.18.0.2): 56 data bytes
    64 bytes from 172.18.0.2: seq=0 ttl=64 time=0.152 ms
    64 bytes from 172.18.0.2: seq=1 ttl=64 time=0.303 ms

    --- db1 ping statistics ---
    2 packets transmitted, 2 packets received, 0% packet loss
    round-trip min/avg/max = 0.152/0.227/0.303 ms

/ # ping -c 2 55e961ed4577
    PING 55e961ed4577 (172.18.0.2): 56 data bytes
    64 bytes from 172.18.0.2: seq=0 ttl=64 time=0.434 ms
    64 bytes from 172.18.0.2: seq=1 ttl=64 time=0.134 ms

    --- 55e961ed4577 ping statistics ---
    2 packets transmitted, 2 packets received, 0% packet loss
    round-trip min/avg/max = 0.134/0.284/0.434 ms

    / # exit
```

在默认网络中创建容器，测试与 my-network 中容器的连通性，示例命令如下：

```
11 $ docker run --rm -itd --name outsider alpine sh

12 $ docker exec -it outsider sh

    / # ip a
    1: lo: <LOOPBACK,UP,LOWER_UP> mtu 65536 qdisc noqueue state UNKNOWN qlen 1000
        link/loopback 00:00:00:00:00:00 brd 00:00:00:00:00:00
        inet 127.0.0.1/8 scope host lo
            valid_lft forever preferred_lft forever
    17: eth0@if18: <BROADCAST,MULTICAST,UP,LOWER_UP,M-DOWN> mtu 1500 qdisc noqueue state UP
        link/ether 02:42:ac:11:00:02 brd ff:ff:ff:ff:ff:ff
        inet 172.17.0.2/16 brd 172.17.255.255 scope global eth0
            valid_lft forever preferred_lft forever

    / # cat /etc/hosts
    127.0.0.1 localhost
    ::1 localhost ip6-localhost ip6-loopback
    fe00::0 ip6-localnet
    ff00::0 ip6-mcastprefix
    ff02::1 ip6-allnodes
    ff02::2 ip6-allrouters
    172.17.0.2    1f25b2d1ffa3
```

```
/ # cat /etc/resolv.conf
# Generated by NetworkManager
nameserver 8.8.8.8
nameserver 114.114.114.114

/ # ping -c 2 db1
ping: bad address 'db1'

/ # ping -c 2 172.18.0.2
PING 172.18.0.2 (172.18.0.2): 56 data bytes

--- 172.18.0.2 ping statistics ---
2 packets transmitted, 0 packets received, 100% packet loss

/ # exit
```

Docker 网络之间的隔离通过 iptables 规则来实现，容器 outsider 默认是无法访问 my-network 网络中的容器的。

10.6.4 接入网络

使用 connect 子命令可以将容器连接到指定网络，它的语法格式如下所示：

```
docker network connect [OPTIONS] NETWORK CONTAINER
```

选项：

- --alias：为容器添加网络级别的别名。
- --driver-opt：网络的驱动程序选项。
- --ip：IPv4 地址。
- --ip6：IPv6 地址。
- --link：添加到其他容器的链接。
- --link-local-ip：为容器添加链接到本地的地址。

可以将容器连接到多个网络，从而实现“多宿主”容器，如图 10-5 所示。

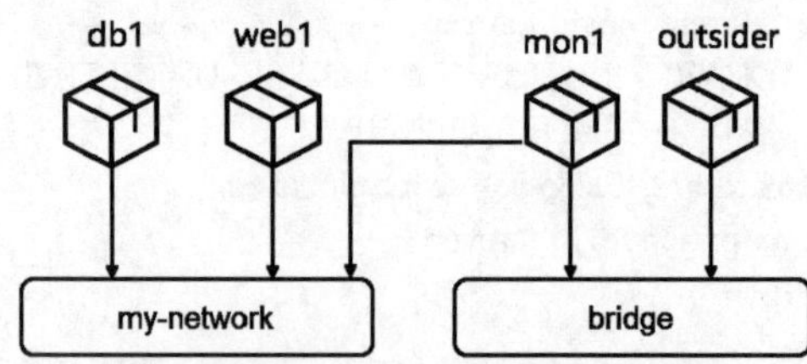

图 10-5 使用自定义网络提高安全性

创建属于网络 my-network 的名为 mon1 的容器，示例命令如下：

```
1 $ docker run --rm -itd --name mon1 --network my-network alpine sh
```

使用 connect 子命令将它也连接到默认的 bridge 网络，示例命令如下：

```
2 $ docker network connect bridge mon1
```

```
3 $ docker ps -a
  CONTAINER ID    IMAGE    COMMAND  CREATED            STATUS              PORTS   NAMES
  405ca56e776c    alpine   "sh"     44 seconds ago     Up 43 seconds        mon1
  70c447455a1a    alpine   "sh"     2 minutes ago      Up 2 minutes        outsider
  04af941dc863    alpine   "sh"     About an hour ago  Up About an hour    web1
  55e961ed4577    alpine   "sh"     About an hour ago  Up About an hour    db1
```

使用 inspect 子命令查看容器 mon1 的网络配置，示例命令如下：

```
4 $ docker container inspect mon1
    ……
    "Networks": {
        "bridge": {
    ……
            "IPAddress": "172.17.0.3",
            "IPPrefixLen": 16,
      ……
        },
        "my-network": {
    ……
            "IPAddress": "172.18.0.4",
            "IPPrefixLen": 16,
    ……
        }
    ……
```

附加到容器 mon1 的控制台，验证网络的连通性，示例命令如下：

```
5 $ docker exec -it mon1 sh

  / # ip a
  1: lo: <LOOPBACK,UP,LOWER_UP> mtu 65536 qdisc noqueue state UNKNOWN qlen 1000
      link/loopback 00:00:00:00:00:00 brd 00:00:00:00:00:00
      inet 127.0.0.1/8 scope host lo
         valid_lft forever preferred_lft forever
  21: eth0@if22: <BROADCAST,MULTICAST,UP,LOWER_UP,M-DOWN> mtu 1500 qdisc noqueue state UP
      link/ether 02:42:ac:12:00:04 brd ff:ff:ff:ff:ff:ff
      inet 172.18.0.4/16 brd 172.18.255.255 scope global eth0
         valid_lft forever preferred_lft forever
  23: eth1@if24: <BROADCAST,MULTICAST,UP,LOWER_UP,M-DOWN> mtu 1500 qdisc noqueue state UP
      link/ether 02:42:ac:11:00:03 brd ff:ff:ff:ff:ff:ff
      inet 172.17.0.3/16 brd 172.17.255.255 scope global eth1
         valid_lft forever preferred_lft forever
```

查看解析器的设置，示例命令如下：

```
/ # cat /etc/hosts
127.0.0.1       localhost
::1     localhost ip6-localhost ip6-loopback
fe00::0 ip6-localnet
ff00::0 ip6-mcastprefix
ff02::1 ip6-allnodes
```

```
ff02::2 ip6-allrouters
172.18.0.4          405ca56e776c
172.17.0.3          405ca56e776c

/ # cat /etc/resolv.conf
nameserver 127.0.0.11
options ndots:0
```

测试与外部网络的连通性，示例命令如下：

```
/ # ip r
default via 172.17.0.1 dev eth1
172.17.0.0/16 dev eth1 scope link   src 172.17.0.3
172.18.0.0/16 dev eth0 scope link   src 172.18.0.4

/ # ping -c 2 www.baidu.com
PING www.baidu.com (220.181.38.149): 56 data bytes
64 bytes from 220.181.38.149: seq=0 ttl=127 time=24.422 ms
64 bytes from 220.181.38.149: seq=1 ttl=127 time=24.546 ms

--- www.baidu.com ping statistics ---
2 packets transmitted, 2 packets received, 0% packet loss
round-trip min/avg/max = 24.422/24.484/24.546 ms
```

测试与 my-network 网络中的容器的连通性，示例命令如下：

```
/ # ping -c 2 db1
PING db1 (172.18.0.2): 56 data bytes
64 bytes from 172.18.0.2: seq=0 ttl=64 time=0.430 ms
64 bytes from 172.18.0.2: seq=1 ttl=64 time=0.138 ms

--- db1 ping statistics ---
2 packets transmitted, 2 packets received, 0% packet loss
round-trip min/avg/max = 0.138/0.284/0.430 ms
```

测试与默认 bridge 网络中的容器的连通性，示例命令如下：

```
/ # ping -c 2 172.17.0.2
PING 172.17.0.2 (172.17.0.2): 56 data bytes
64 bytes from 172.17.0.2: seq=0 ttl=64 time=0.163 ms
64 bytes from 172.17.0.2: seq=1 ttl=64 time=0.178 ms

--- 172.17.0.2 ping statistics ---
2 packets transmitted, 2 packets received, 0% packet loss
round-trip min/avg/max = 0.163/0.170/0.178 ms
```

10.6.5 断开网络

使用 disconnect 子命令可以将容器从网络上断开，它的语法格式如下：

```
docker network disconnect [OPTIONS] NETWORK CONTAINER
```

选项：

- -f, --force：强制把容器从网络上断开。

创建新的容器再连接到两个网络，然后断开其中一个网络，示例命令如下：

```
1 $ docker network ls
    NETWORK ID      NAME          DRIVER     SCOPE
    a98eb23887e0    bridge        bridge     local
    84805310515b    host          host       local
    ec5b7b2ebf70    my-network    bridge     local
    e714a6c02456    none          null       local

2 $ docker run --rm -d --name mon1 alpine sleep 1d

3 $ docker exec mon1 ip a
    1: lo: <LOOPBACK,UP,LOWER_UP> mtu 65536 qdisc noqueue state UNKNOWN qlen 1000
        link/loopback 00:00:00:00:00:00 brd 00:00:00:00:00:00
        inet 127.0.0.1/8 scope host lo
            valid_lft forever preferred_lft forever
    5: eth0@if6: <BROADCAST,MULTICAST,UP,LOWER_UP,M-DOWN> mtu 1500 qdisc noqueue state UP
        link/ether 02:42:ac:11:00:02 brd ff:ff:ff:ff:ff:ff
        inet 172.17.0.2/16 brd 172.17.255.255 scope global eth0
            valid_lft forever preferred_lft forever
```

第 3 行命令的输出显示：如果创建容器时未指定网络，则会连接到默认的 bridge 网络。

```
4 $ docker network connect my-network mon1

5 $ docker exec mon1 ip a
    1: lo: <LOOPBACK,UP,LOWER_UP> mtu 65536 qdisc noqueue state UNKNOWN qlen 1000
        link/loopback 00:00:00:00:00:00 brd 00:00:00:00:00:00
        inet 127.0.0.1/8 scope host lo
            valid_lft forever preferred_lft forever
    5: eth0@if6: <BROADCAST,MULTICAST,UP,LOWER_UP,M-DOWN> mtu 1500 qdisc noqueue state UP
        link/ether 02:42:ac:11:00:02 brd ff:ff:ff:ff:ff:ff
        inet 172.17.0.2/16 brd 172.17.255.255 scope global eth0
            valid_lft forever preferred_lft forever
    7: eth1@if8: <BROADCAST,MULTICAST,UP,LOWER_UP,M-DOWN> mtu 1500 qdisc noqueue state UP
        link/ether 02:42:ac:12:00:02 brd ff:ff:ff:ff:ff:ff
        inet 172.18.0.2/16 brd 172.18.255.255 scope global eth1
            valid_lft forever preferred_lft forever
```

第 5 行命令的输出显示：将容器连接到指定网络，会立即生效。

```
6 $ docker network disconnect my-network mon1

7 $ docker exec mon1 ip a
    1: lo: <LOOPBACK,UP,LOWER_UP> mtu 65536 qdisc noqueue state UNKNOWN qlen 1000
        link/loopback 00:00:00:00:00:00 brd 00:00:00:00:00:00
        inet 127.0.0.1/8 scope host lo
            valid_lft forever preferred_lft forever
    5: eth0@if6: <BROADCAST,MULTICAST,UP,LOWER_UP,M-DOWN> mtu 1500 qdisc noqueue state UP
        link/ether 02:42:ac:11:00:02 brd ff:ff:ff:ff:ff:ff
        inet 172.17.0.2/16 brd 172.17.255.255 scope global eth0
```

```
        valid_lft forever preferred_lft forever
```

第 7 行命令的输出显示：将容器从指定网络断开，也会立即生效。

10.6.6 删除和清理网络

当没有容器连接到网络时，可以使用 rm 子命令将这个网络删除，它的语法格式如下：

```
docker network rm NETWORK [NETWORK...]
```

也可以使用 prune 子命令清理无容器使用的自定义网络，它的语法格式如下：

```
docker network prune [OPTIONS]
```

选项：

- --filter：设置显示的过滤条件，例如：until=<timestamp>。
- -f, --force：不需要确认操作。

创建新的容器再连接到自定义网络，然后进行删除和清理操作，示例命令如下：

```
1 $ docker network ls
   NETWORK ID      NAME          DRIVER      SCOPE
   a98eb23887e0    bridge        bridge      local
   84805310515b    host          host        local
   ec5b7b2ebf70    my-network    bridge      local
   e714a6c02456    none          null        local

2 $ docker run --rm -d --name a1 --network my-network alpine sleep 1d

3 $ docker network rm my-network
   Error response from daemon: error while removing network: network my-network id
   ec5b7b2ebf7065562f9f1c43972b5029e18562a42288137a960717df3541e705 has active endpoints
```

由于还有容器连接到 my-network 网络，所以执行第 3 行命令会出错。

```
4 $ docker stop a1

5 $ docker network inspect my-network
   [
       {
           "Name": "my-network",
           "Id": "ec5b7b2ebf7065562f9f1c43972b5029e18562a42288137a960717df3541e705",
           "Created": "2022-12-12T20:18:54.317661119+08:00",
           "Scope": "local",
           "Driver": "bridge",
           "EnableIPv6": false,
           "IPAM": {
               "Driver": "default",
               "Options": {},
               "Config": [
                   {
                       "Subnet": "172.18.0.0/16",
                       "Gateway": "172.18.0.1"
```

```
                    }
                ]
            },
            "Internal": false,
            "Attachable": false,
            "Ingress": false,
            "ConfigFrom": {
                "Network": ""
            },
            "ConfigOnly": false,
            "Containers": {},
            "Options": {},
            "Labels": {}
        }
    ]


6 $ docker network prune
    WARNING! This will remove all custom networks not used by at least one container.
    Are you sure you want to continue? [y/N] y
    Deleted Networks:
    my-network

7 $ docker network ls
    NETWORK ID     NAME      DRIVER    SCOPE
    a98eb23887e0   bridge    bridge    local
    84805310515b   host      host      local
    e714a6c02456   none      null      local
```

从第 5 行命令的输出中可以看出：没有容器连接到自定义网络 my-network，所以可以安全地进行清除操作。

10.7 配置 host 网络模式

在 bridge 网络模式中，容器与宿主机网络之间是隔离的。而在 host 网络模式中，是没有这种隔离的，容器直接连接到宿主机的网络，如图 10-6 所示。

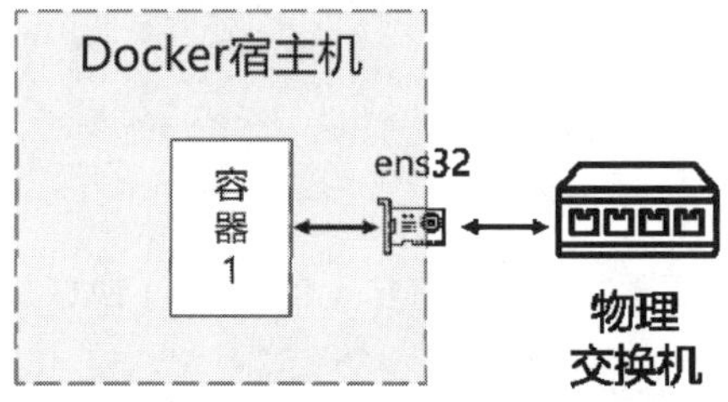

图 10-6　host 类型的容器网络

如果容器使用 host 网络模式，则该容器的网络堆栈没有与宿主机隔离，而是共享宿主机的网络，并且该容器没有 IP 和 MAC 地址。

在后台运行一个 host 网络模式的 httpd 容器，会直接将其绑定到宿主机的 80 端口，示例命令如下：

```
1 $ docker run --rm -d --network host --name web1 httpd

2 $ docker ps -a
    CONTAINER ID  IMAGE    COMMAND              CREATED         STATUS         PORTS    NAMES
    f4a7fc6a6c1b  httpd    "httpd-foreground"   5 seconds ago   Up 5 seconds            web1

3 $ docker network inspect host
    [
        {
            "Name": "host",
            "Id": "84805310515be4c9d2ba3cb5dcbc15fa182a78f63876e60681b74a4a6cba8a3a",
            "Created": "2022-07-19T15:26:34.819232394+08:00",
            "Scope": "local",
            "Driver": "host",
            "EnableIPv6": false,
            "IPAM": {
                "Driver": "default",
                "Options": null,
                "Config": []
            },
            "Internal": false,
            "Attachable": false,
            "Ingress": false,
            "ConfigFrom": {
                "Network": ""
            },
            "ConfigOnly": false,
            "Containers": {
                "f4a7fc6a6c1b2e627ab89c6bb6ba8554b5ff21020712d1e719afed9dae36d33a": {
                    "Name": "web1",
                    "EndpointID": "0bb348d714032688b43fe87dcd9e7154b92affe9e3fc59493a7582ec66a3a6e3",
                    "MacAddress": "",
                    "IPv4Address": "",
                    "IPv6Address": ""
                }
            },
            "Options": {},
            "Labels": {}
        }
    ]
```

第 3 行命令的输出显示：容器 web1 没有自己的 IP 和 MAC 地址。

```
4 $ sudo ss -tulpn | grep :80
    tcp    LISTEN 0      511          *:80          *:*        users:(("httpd",pid=2958,fd=4),("
httpd",pid=2778,fd=4),("httpd",pid=2777,fd=4),("httpd",pid=2776,fd=4),("httpd",pid=2765,fd=4))
```

第 4 行命令的输出显示：httpd 进程直接运行在宿主机上，它监听 TCP 80 端口，而不是在容器中运行隔离级别的进程。

```
5 $ curl localhost
    <html><body><h1>It works!</h1></body></html>
```

第 5 行命令可以正确获得容器中的 Web 页面。如果需要从宿主机外部网络访问，还需要配置防火墙，示例命令如下：

```
6 $ sudo firewall-cmd --add-service=http
```

容器使用 host 网络模式的场景非常有限，主要是为了提高网络性能。因为它不需要网络地址转换(NAT)，所以性能更好一些。

10.8 配置 none 网络模式

如果容器使用的是 none 网络模式，则会禁用所有网络。

```
1 $ docker run -it --rm --name a1 --network none alpine ip a
    1: lo: <LOOPBACK,UP,LOWER_UP> mtu 65536 qdisc noqueue state UNKNOWN qlen 1000
        link/loopback 00:00:00:00:00:00 brd 00:00:00:00:00:00
        inet 127.0.0.1/8 scope host lo
           valid_lft forever preferred_lft forever


2 $ docker run -it --rm --name a1 --network none alpine mkpasswd
    Password:回车
$6$nQyDjA1nHkbtDNcg$EOACj6u4tvlJc2c9kOLomiluwhUcE6dYb4eaxvkJBuGvIZomovU0e6WC5sf961EdL
  qNMwxi3N9UTd/FRGUWDo0
```

第 1 行命令的输出显示：采用 none 网络模式的容器只有 loop 接口。

这种类型的容器既不能访问外部网络，也不能与其他容器通信，适用于需要禁用网络以确保安全性等特殊的应用场景，如第 2 行命令所示。

10.9 习题

1. [多选题]docker 的网络模式包括(　　)。

   A. bridge　　B. host

   C. container　　D. none

2. [单选题]Docker 容器网络拥有自己的(　　)。

   A. 容器　　B. 储存空间

   C. 网络命名空间　　D. 数据空间

3. [单选题]选项(　　)将创建一个 Nginx 容器并加入指定网络 lnmp_net。

A. docker run -net lnmp_net nginx

B. docker run -n nmp_net nginx

C. docker run --network lnmp_net nginx

D. docker run --attach lnmp_net nginx

5. [单选题]下列关于 Docker 容器互联的说法，错误的是(　　)。

A. 容器互联是在容器间建立的一条专门的网络通信隧道

B. 容器互联是通过容器的名称实现的

C. 容器互联后，源容器可以看到接收容器指定的信息

D. 可以避免暴露端口到外部网络，提高容器的安全性

6. [单选题]关于 Docker 端口映射，正确的是(　　)。

A. 创建容器的时候，只可使用一次-p 选项，一个容器绑定一个端口

B. 映射 udp 端口的时候，可以采用的方式为-p 5000:5000/udp

C. -p 4000:5000 表示将宿主机上的 5000 端口映射到容器的 4000 端口

D. 可以使用 docker logs 查看具体的端口映射情况

7. [单选题]下列关于 Docker 端口映射的描述，不正确的是(　　)。

A. 容器不进行端口映射，可以通过网络访问容器内的服务

B. 端口映射就是将宿主机的端口映射到容器中

C. 使用-P(大写)实现端口映射时，需要关注镜像对外暴露的端口

D. 使用-p(小写)可以实现端口映射

9. [单选题]容器网络模式指定(　　)与已存在的容器共享同一个网络命名空间。

A. 新容器　　　　B. 宿主机

C. 其他宿主机中的容器　　　　D. 原容器

10. [判断题]Docker 目前不推荐使用-link 的方式进行容器互联，而是建议建立同一自定义网络的容器互联的方式。这种说法(　　)。

A. 正确　　　　B. 错误

第 11 章

Docker存储管理

Docker 存储分为持久化存储与非持久化存储。通常不需要对非持久化存储进行管理，管理的重点是持久化存储。

本章要点

◎ Docker 存储概述

◎ Docker 的卷(volume)

◎ Docker 的绑定挂载(bind mount)

11.1 Docker 存储概述

非持久化数据，就是不需要一直保存的数据。组成 Docker 镜像的各层均是只读的，运行容器的时候，会在其上附加一个读写层，构建出统一的文件系统，这就是容器的非持久化存储。非持久化存储是自动创建的，生命周期也与容器相同。删除容器，也会删除这些非持久化数据。

持久化数据就是需要永久保存的数据，例如，客户信息、财务凭证、审计日志等数据。持久化数据与容器是解耦的，未与任何容器生命周期绑定，可以独立地创建并管理它们。即使删除容器，也不会删除这些数据。

Docker 有两种常用的实现持久化数据的方法：卷(volume)和绑定挂载(bind mount)。如果使用卷，需要先创建 Docker 卷(手工或自动创建)，然后将其挂载到容器的某个目录中(例如：/code)，如图 11-1 所示。任何保存到该目录下的内容都会被写到卷中，即使容器被删除，卷与其中的数据仍然存在。

Docker 还支持在宿主机内存中存储文件(tmpfs 挂载)，此类数据不会持久化。

无论选择使用哪种类型的挂载，容器中的数据看起来都一样，它们在容器的文件系统中显示为目录或单个文件。它们之间的主要区别是在 Docker 宿主机上的位置，如图 11-2 所示。

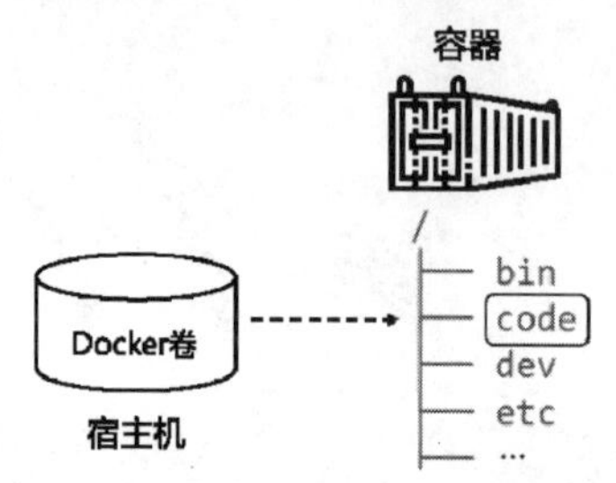

图 11-1　挂载 Docker 卷到容器的/code 目录

图 11-2　Docker 的存储类型

(1) 卷：由 Docker 管理的宿主机文件系统的一部分，默认位置是/var/lib/Docker/Volumes/。卷是在 Docker 中保存数据的最佳方式，强烈不建议非 Docker 进程修改这部分数据。

(2) 绑定挂载：可以是宿主机系统的任何位置，甚至可能是重要的系统文件或目录。除了容器之外，宿主机上非 Docker 进程也可以随时修改其中的文件。

(3) tmpfs 挂载：只使用宿主机系统的内存。

11.2 Docker 的卷

卷(volume)是保存 Docker 容器生成和使用数据的首选机制，它是 Docker 管理的宿主机中的一个目录，可以将一个卷同时挂载到多个容器中。

通过 Docker 的 volume 子命令来管理卷，volume 子命令还包括以下几个子命令：

- create：创建卷。
- inspect：显示一个或多个卷的详细信息。
- ls：列出卷。
- prune：删除所有未使用的本地卷。
- rm：删除一个或多个卷。

11.2.1 Docker 卷的管理

根据 Docker 卷名称的由来，可以将卷分为两类：

- 命名(named)卷：在使用卷之前创建的而且指定卷名称的卷。
- 匿名(anonymous)卷：随创建容器一起创建的卷，或者创建卷时未指定名称的卷，Docker 会生成一个唯一性的随机名称。

除了名称之外，命名卷和匿名卷没有什么区别。

使用 create 子命令创建卷，它的语法格式如下：

```
docker volume create [OPTIONS] [VOLUME]
```

选项：

- -d, --driver：指定卷驱动程序，默认为 local。

- --label：设置卷的元数据。
- -o, --opt：设置驱动程序的特定选项。

创建卷的时候可以使用-d 选项指定驱动程序，默认为 local 驱动，用于构建本地卷。本地卷只能被所在节点上的容器使用。Docker 还可以通过插件的方式使用第三方存储系统(包括块存储、文件存储、对象存储等)，充分利用它们的高级特性，如图 11-3 所示。

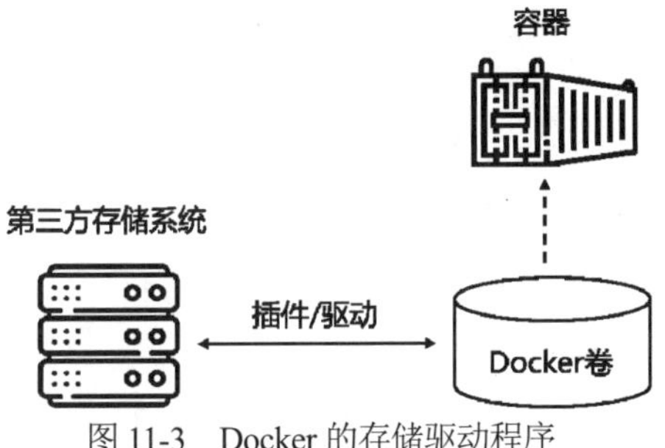

图 11-3　Docker 的存储驱动程序

创建一个名为 myvol 的本地卷，示例命令如下：

```
1 $ docker volume create myvol

2 $ docker volume ls
    DRIVER          VOLUME NAME
    local           myvol

3 $ docker volume inspect myvol
    [
        {
            "CreatedAt": "2022-12-28T20:57:20+08:00",
            "Driver": "local",
            "Labels": {},
            "Mountpoint": "/var/lib/docker/volumes/myvol/_data",
            "Name": "myvol",
            "Options": {},
            "Scope": "local"
        }
    ]

4 $ sudo tree /var/lib/docker/volumes/
    /var/lib/docker/volumes/
    ├── backingFsBlockDev
    ├── metadata.db
    └── myvol
        └── _data
```

创建卷之后，使用 ls 子命令查看宿主机上所有卷的列表，使用 inspect 子命令查看指定卷的详细信息，通过 Mountpoint 属性可以获得卷所在的目录。

11.2.2 Docker 卷的使用

创建卷之后，就可以在 docker container run 命令中通过--mount 选项将其挂载到新建容器中，语法格式如下：

```
--mount type=TYPE,TYPE-SPECIFIC-OPTION[,...]
```

主要选项如下：

- type：当前支持的挂载类型有 bind、volume 和 tmpfs。
- src, source：指定挂载源。
- dst, destination, target：指定挂载目标。
- ro, readonly：指定是否为只读挂载，默认为 false。

语法示例如下：

```
type=bind,source=/path/on/host,destination=/path/in/container

type=volume,source=my-volume,destination=/path/in/container,volume-label="color=red",volume-label="shape=
    round"

type=tmpfs,tmpfs-size=512M,destination=/path/in/container
```

创建容器时使用现有的卷，示例命令如下：

```
1 $ docker run --rm -it --name app1 --mount source=myvol,target=/data alpine sh
    / # mount | grep data
    /dev/mapper/cs-root on /data type xfs (rw,seclabel,relatime,attr2,inode64,logbufs=8,logbsize=32k,noquota)

    / # date > /data/1.txt
Ctrl+p、Ctrl+q 断开会话
```

第 1 行命令中的 source 选项指定源，也就是卷的名称。target 选项指定容器目录树中的路径，如果没有这个目录会自动创建，这个目录必须是绝对路径。容器中的这个目录中创建用于后续测试的文件。

```
2 $ docker run --rm -it --name app2 --mount source=myvol,target=/data alpine sh
    / # cat /data/1.txt
    Tue Dec 20 12:48:59 UTC 2022

    / # date > /data/2.txt
Ctrl+p、Ctrl+q 断开会话
```

第 2 行命令再创建一个新的容器，也挂载这个卷。

```
3 $ sudo ls /var/lib/docker/volumes/myvol/_data/
    1.txt    2.txt

4 $ echo 333 | sudo tee /var/lib/docker/volumes/myvol/_data/3.txt
```

在新容器中，既可以访问卷中原有的文件，也可以创建新的文件。

```
5 $ sudo ls -li /var/lib/docker/volumes/myvol/_data/
    total 12
```

```
    201665076 -rw-r--r--. 1 root root 29 Dec 20 20:48 1.txt
    201665084 -rw-r--r--. 1 root root 29 Dec 20 20:51 2.txt
    201665085 -rw-r--r--. 1 root root 4 Dec 20 20:58 3.txt

6 $ docker exec -it app1 ls -li /data
    total 12
    201665076 -rw-r--r--    1 root root 29 Dec 20 12:48 1.txt
    201665084 -rw-r--r--    1 root root 29 Dec 20 12:51 2.txt
    201665085 -rw-r--r--    1 root root 4 Dec 20 12:58 3.txt

7 $ docker exec -it app2 ls -li /data
    total 12
    201665076 -rw-r--r--    1 root root 29 Dec 20 12:48 1.txt
    201665084 -rw-r--r--    1 root root 29 Dec 20 12:51 2.txt
    201665085 -rw-r--r--    1 root root 4 Dec 20 12:58 3.txt
```

从第 3~第 7 行的命令输出可以看出，卷可以在宿主机、多个容器之间进行共享访问。

11.2.3 Docker 卷的适用场景

卷是在 Docker 容器和服务中保存持久化数据的首选方式，它的适用场景包括：

- 在多个正在运行的容器之间共享数据。如果没有手工显式地创建卷，将会在第一次使用时自动创建卷。当停止或删除容器时，卷仍然存在。只有当显式明确地删除卷时，才会删除它们。示例命令如下：

```
$ docker volume ls
  DRIVER VOLUME NAME

$ docker run --rm -itd --name test1 --mount source=mydata,target=/data alpine sleep 1d

$ docker volume ls
  DRIVER  VOLUME NAME
  local mydata
```

- 当宿主机不能保证具有给定的目录或文件结构时，通过卷可以很方便地实现存储与容器的解耦。
- 当需要从一台宿主机向另一台宿主机备份、恢复或迁移数据时，卷是更好的选择。
- 当希望将容器的数据存储在远程宿主机或云(而不是本地)上时，卷是很好的选择。
- 在 Docker Desktop 版本，使用卷会更加灵活一些。

11.3 Docker 的绑定挂载

绑定挂载(bind mount)是 Docker 早期的技术，功能有限。但它的优点是可以将宿主机上的任何文件或目录(需要使用绝对路径)挂载到容器中。Docker 的卷只能使用/var/lib/docker/volumes/目录，而绑定挂载则可以使用宿主机的整体文件系统，甚至可以使用宿主机上的敏感文件。这些被挂载的文件或目录可以被任何进程修改，包括容器中的进程、宿主机的进程，所以要关注它们之间是否相互影响。

绑定挂载的主要应用场景包括：

- 从宿主机共享配置文件到容器。默认情况下，Docker 会绑定/etc/resolv.conf 等的文件用于 DNS 的解析。
- 宿主机与容器共享源代码或构建工具。
- 容器需要与宿主机的文件或目录结构一致时。

在使用绑定挂载时，需要注意如下特性：

- 如果将宿主机的非空目录挂载给容器，而且容器中的该目录也有文件，那么容器中的文件将会被隐藏。
- 如果将宿主机的空目录挂载给容器，而且容器中的该目录有文件，这些文件将会被复制到宿主主机上的目录中。

实现绑定挂载，既可以使用--mount 选项(需要指定 type=bind)，也可以使用-v，--volume 选项。从实现效果来看，两者没有区别。

使用--mount 选项的示例命令如下：

```
$ docker run -d   -it   --name devtest \
  --mount type=bind,source="$(pwd)"/target,target=/app nginx:latest
```

使用-v，--volume 选项，语法格式如下所示：

```
-v|--volume[=[[HOST-DIR:]CONTAINER-DIR[:OPTIONS]]]
```

如果指定-v /HOST-DIR:/CONTAINER-DIR，Docker 将宿主机中的/HOST-DIR 挂载到 Docker 容器中的/CONTAINER-DIR。如果未指定 HOST-DIR，Docker 会自动在宿主机上创建新目录。OPTIONS 是由逗号分隔的选项列表。示例命令如下：

```
$ docker run -d    -it --name devtest \
  -v "$(pwd)"/target:/app   nginx:latest
```

不管是哪种方式创建的绑定挂载，在容器层面都是一样的。示例命令如下：

```
$ docker inspect devtest
  ……
  "Mounts": [
      {
          "Type": "bind",
          "Source": "/tmp/source/target",
          "Destination": "/app",
          "Mode": "",
          "RW": true,
          "Propagation": "rprivate"
      }
  ],
  ……
```

接下来通过绑定挂载给 Nginx 容器提供持久数据存储。首先考察 Nginx 容器的默认配置，示例命令如下：

```
1 $ docker run --rm -d --name web1 -P nginx

2 $ docker ps -a
    CONTAINER ID IMAGE COMMAND                CREATED        STATUS
    70cf0ebe0dcd    nginx    "/docker-entrypoint.…"  5 seconds ago   Up 3 seconds
PORTS     NAMES
0.0.0.0:49153->80/tcp,   :::49153->80/tcp     web1

3 $ curl localhost:49153
    <!DOCTYPE html>
    <html>
    <head>
    <title>Welcome to nginx!</title>
    ……

4 $ docker exec -it web1 cat /etc/nginx/conf.d/default.conf | grep root
            root     /usr/share/nginx/html;
            root     /usr/share/nginx/html;
        # root html;
        # deny access to .htaccess files, if Apache's document root

5 $ docker stop web1
```

通过第 4 行命令获得网站的根目录是/usr/share/nginx/html。

```
6 $ docker run --rm -d --name web2 -v /web:/usr/share/nginx/html -P nginx
```

第 6 行命令中-v 选项将宿主机本地目录/web 挂载到容器的网站根目录，如果宿主机本地没有指定的目录，会自动创建目录。

```
7 $ docker ps
    CONTAINER ID  IMAGE    COMMAND                  CREATED        STATUS
    174642726c6a     nginx      "/docker-entrypoint.…"    5 seconds ago   Up 4 seconds
    PORTS  NAMES
    0.0.0.0:49154->80/tcp,        :::49154->80/tcp      web2

8 $ echo "Test web site" | sudo tee /web/index.html

9 $ curl localhost:49154
Test web site
```

如果开发人员修改了宿主机本地的页面文件，会在网站上看到变化，示例命令如下：

```
10 $ echo "2222" | sudo tee -a /web/index.html

11 $ curl localhost:49154
    Test web site
    2222
```

11.4 习题

1. [单选题]可以查看正在使用的存储驱动程序的命令是(　　)。

   A. docker volume　　B. docker info

   C. docker hint　　D. docker pretty

2. [单选题]恢复数据卷是将备份数据恢复到(　　)中。

   A. 新容器　　B. 数据卷容器

   C. 原容器　　D. Web 容器

3. [单选题]数据卷最大的优势是可以用来做(　　)数据。

   A. 储存　　B. 业务

   C. 持久化　　D. 数据库

4. [单选题]Docker 的数据管理是通过(　　)实现的。

   A. lvm　　B. 端口映射

   C. 镜像　　D. 数据卷

5. [单选题]以下关于容器数据卷的说法，错误的是(　　)。

   A. 数据卷不可以在容器之间共享和重用

   B. 对数据卷的修改会立马生效

   C. 对数据卷的更新不会影响镜像

   D. 当容器被销毁时，容器所使用的数据卷不会被删除

第 12 章

使用Dockerfile创建镜像

Dockerfile 是在自动创建镜像时使用的文本格式的配置文件，使用它是首选的构建镜像的方式。

本章要点

◎ Dockerfile 的基本结构
◎ Dockerfile 的配置指令
◎ Dockerfile 的操作指令
◎ 创建镜像

12.1 Dockerfile 的基本结构

Dockerfile 由多行指令语句组成，包括“配置指令”(配置镜像信息)和“操作指令”(具体执行操作)。以#开头的为注释行。Dockerfile 主体包括：

- 基础镜像信息。
- 维护者信息。
- 镜像操作指令。
- 容器启动时的执行指令。

Dockerfile 可以很简单，例如 Alpine Linux 容器的 Dockerfile 只有 3 行，如图 12-1 所示。也可以很复杂，例如 Nginx 容器的 Dockerfile 就多达 114 行，如图 12-2 所示。

```
3 lines (3 sloc) | 76 Bytes
1 FROM scratch
2 ADD alpine-minirootfs-20220715-x86_64.tar.gz /
3 CMD ["/bin/sh"]
```

图 12-1　Alpine Linux 镜像的 Dockerfile 文件

```
114 lines (107 sloc) | 5.26 KB
#
# NOTE: THIS DOCKERFILE IS GENERATED VIA "update.sh"
#
# PLEASE DO NOT EDIT IT DIRECTLY.
#
FROM debian:bullseye-slim

LABEL maintainer="NGINX Docker Maintainers <docker-maint@nginx.com>"

ENV NGINX_VERSION   1.23.1
ENV NJS_VERSION     0.7.6
ENV PKG_RELEASE     1~bullseye

RUN set -x \
# create nginx user/group first, to be consistent throughout docker variants
    && addgroup --system --gid 101 nginx \
    && adduser --system --disabled-login --ingroup nginx --no-create-home --h
```

图 12-2　Nginx 镜像的 Dockerfile 文件

Docker 由许多指令组成，在创建 Docker 容器时，Docker 中的每个指令代表一层：

- ADD：向镜像中添加文件，支持压缩格式。
- COPY：复制文件到镜像。
- CMD：在容器内执行的特定命令。
- ENTRYPOINT：设置了默认应用程序，以便在每次使用镜像创建容器时使用。
- ENV：设置环境变量。
- EXPOSE：设置暴露的端口。
- FROM：定义基础镜像。
- MAINTAINER：定义创建者的信息。
- RUN：运行指定的指令。
- USER：指定运行容器时的用户名或 UID。
- VOLUME：创建一个数据卷挂载点。
- WORKDIR：为后续的 RUN、CMD、ENTRYPOINT 指令配置工作目录。
- LABEL：向镜像添加自定义标签。

构建镜像的 Dockerfile 并不需要使用所有的关键字，例如，Alpine Linux 镜像的 Dockerfile 如下：

```
FROM scratch
ADD alpine-minirootfs-20220715-x86_64.tar.gz /
CMD ["/bin/sh"]
```

第 1 行 FROM 指令，表示此镜像使用 scratch 作为基础镜像。

第 2 行 ADD 指令，将 alpine-minirootfs-20220715-x86_64.tar.gz 文件添加至镜像中，并解压缩到根目录。

第 3 行 CMD 指令，设置当容器启动时，运行"/bin/sh"命令。

查看 Alpine Linux 镜像的构建历史记录，示例命令如下：

```
$ docker history alpine
  IMAGE          CREATED       CREATED BY                                    SIZE      COMMENT
  d7d3d98c851f   9 days ago    /bin/sh -c #(nop)   CMD ["/bin/sh"]                     0B
  <missing>      9 days ago    /bin/sh -c #(nop) ADD file:a2648378045615c37…            5.53MB
```

构建这个镜像分为 3 个步骤。第 1 步根据基础镜像启动新容器，scratch 是一个特殊的基础镜像(不包括任何数据的空白层)。第 2 步由 ADD 指令添加 TAR 文件，这将创建一个中间的临时容器。第 3 步由 CMD 指令再添加一个新层构建另一个新容器，之前的临时容器会被删除，所以在镜像构建历史记录中，会看到有<missing>类信息。Alpine Linux 镜像的构建过程如图 12-3 所示。

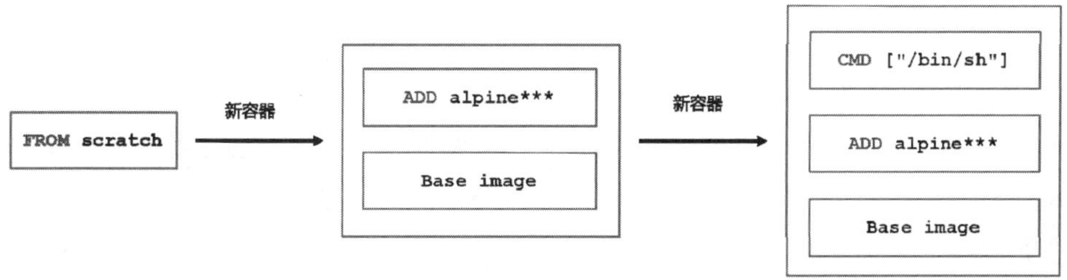

图 12-3　Alpine Linux 镜像的构建过程

Docker Hub 网站会提供每个官方镜像的 Dockerfile 文件，可以通过阅读它们来更好地学习。下面简单分析一下 Nginx 的 Dockerfile 文件。

```
1 #
2 # NOTE: THIS DOCKERFILE IS GENERATED VIA "update.sh"
3 #
4 # PLEASE DO NOT EDIT IT DIRECTLY.
5 #
```

前面这几行是注释信息，提示此 Dockerfile 是通过 update.sh 生成的，请不要直接编辑。可以在这个项目的 Git Hub 网站上找到脚本 update.sh。

```
6 FROM debian:bullseye-slim
```

第 6 行 FROM 指令将 debian:bullseye 作为基础镜像。可以在 Docker Hub 网站上找到 debian:bullseye-slim 的 Dockerfile，它很简单，只有 3 行：

```
FROM scratch
ADD rootfs.tar.xz /
CMD ["bash"]
```

Debian 镜像是从 scratch 镜像创建的，将 rootfs.tar.xz 文件解压缩是从根目录开始的，最后运行 bash 命令。因此在 Nginx 镜像中，最初的 2 层来自基础镜像 Debian。

```
$ docker image history nginx:latest
    IMAGE          CREATED      CREATED BY                                                   SIZE   COMMENT
    670dcc86b69d   9 days ago   /bin/sh -c #(nop)   CMD ["nginx" "-g" "daemon…                0B
    <missing>      9 days ago   /bin/sh -c #(nop)   STOPSIGNAL SIGQUIT                         0B
```

```
<missing>     9 days ago    /bin/sh -c #(nop)  EXPOSE 80                        0B
<missing>     9 days ago    /bin/sh -c #(nop)  ENTRYPOINT ["/docker-entr…       0B
<missing>     9 days ago    /bin/sh -c #(nop)  COPY file:09a214a3e07c919a…      4.61kB
<missing>     9 days ago    /bin/sh -c #(nop)  COPY file:0fd5fca330dcd6a7…      1.04kB
<missing>     9 days ago    /bin/sh -c #(nop)  COPY file:0b866ff3fc1ef5b0…      1.96kB
<missing>     9 days ago    /bin/sh -c #(nop)  COPY file:65504f71f5855ca0…      1.2kB
<missing>     9 days ago    /bin/sh -c set -x  && addgroup --system -…          61.1MB
<missing>     9 days ago    /bin/sh -c #(nop)  ENV PKG_RELEASE=1~bullseye       0B
<missing>     9 days ago    /bin/sh -c #(nop)  ENV NJS_VERSION=0.7.6            0B
<missing>     9 days ago    /bin/sh -c #(nop)  ENV NGINX_VERSION=1.23.1         0B
<missing>     2 weeks ago   /bin/sh -c #(nop)  LABEL maintainer=NGINX Do…       0B
<missing>     2 weeks ago   /bin/sh -c #(nop)  CMD ["bash"]                     0B
<missing>     2 weeks ago   /bin/sh -c #(nop)  ADD file:d978f6d3025a06f51…      80.4MB

7
8 LABEL maintainer="NGINX Docker Maintainers <docker-maint@nginx.com>"
9
```

第 8 行 LABEL 指令为生成的镜像添加元数据标签信息。这些信息可以用来辅助过滤出特定镜像。

```
10 ENV NGINX_VERSION      1.23.3
11 ENV NJS_VERSION        0.7.9
12 ENV PKG_RELEASE        1~bullseye
13
```

第 10~第 12 行 ENV 指令设置了 3 个环境变量。在镜像生成过程中，后续使用 RUN 指令会应用这些环境变量(如第 34~第 38 行)。

提示

在启动容器时，也可以指定环境变量。其格式为 ENV <key><value>或 ENV<key>=<value>...。

第 14~第 102 行其实是一个 RUN 指令，它们是运行多个命令。每个 RUN 指令会在当前镜像基础上执行指定命令，为了减少容器的层数，降低维护的工作量，可以将多个命令通过&&连接在一起。当命令较长时，可以使用\来换行。

```
14 RUN set -x \
```

默认情况下，脚本执行后，屏幕只显示运行结果。如果连续执行多个命令，运行结果就会混在一起。为了方便排错，第 14 行通过 set -x 设置在输出结果之前，先输出要执行的那一行命令。

```
15 # create nginx user/group first, to be consistent throughout docker variants
16      && addgroup --system --gid 101 nginx \
17      && adduser --system --disabled-login --ingroup nginx --no-create-home --home /nonexistent --gecos
           "n nx user" --shell /bin/false --uid 101 nginx \
```

第 15 行是注释，提示说“首先创建 nginx 用户/组，以在整个 docker 变体中保持一致”。

第 16 行的&&是连接两个命令的符号，这行指令将在系统中添加新的用户组 nginx。

第 17 行指令在系统中添加新的用户，也叫 nginx。

```
18      && apt-get update \
19      && apt-get install --no-install-recommends --no-install-suggests -y gnupg1 ca-certificates \
20      && \
```

第 18 行指令更新操作系统。

第 19 行指令安装 gnupg1、ca-certificate 等软件组件。

随后的多行指令，可以认为是类似“一键安装 Nginx”功能的脚本。

```
21      NGINX_GPGKEY=573BFD6B3D8FBC641079A6ABABF5BD827BD9BF62; \
22      found=''; \
23      for server in \
24          hkp://keyserver.ubuntu.com:80 \
25          pgp.mit.edu \
26      ; do \
27          echo "Fetching GPG key $NGINX_GPGKEY from $server"; \
28          apt-key adv --keyserver "$server" --keyserver-options timeout=10 --recv-keys
        "$NGINX_GPGKEY" && fo d=yes && break; \
29      done; \
30      test -z "$found" && echo >&2 "error: failed to fetch GPG key $NGINX_GPGKEY" && exit 1; \
31      apt-get remove --purge --auto-remove -y gnupg1 && rm -rf /var/lib/apt/lists/* \
32      && dpkgArch="$(dpkg --print-architecture)" \
33      && nginxPackages=" \
34          nginx=${NGINX_VERSION}-${PKG_RELEASE} \
35          nginx-module-xslt=${NGINX_VERSION}-${PKG_RELEASE} \
36          nginx-module-geoip=${NGINX_VERSION}-${PKG_RELEASE} \
37          nginx-module-image-filter=${NGINX_VERSION}-${PKG_RELEASE} \
38          nginx-module-njs=${NGINX_VERSION}+${NJS_VERSION}-${PKG_RELEASE} \
39      " \
40      && case "$dpkgArch" in \
41          amd64|arm64) \
42 # arches officialy built by upstream
43              echo "deb https://nginx.org/packages/mainline/debian/ bullseye nginx" >> /etc/apt/sour
             ces.list.d/nginx.list \
44              && apt-get update \
45              ;; \
46          *) \
47 # we're on an architecture upstream doesn't officially build for
48 # let's build binaries from the published source packages
49              echo "deb-src https://nginx.org/packages/mainline/debian/ bullseye nginx" >> /etc/apt/sour
              ces.list.d/nginx.list \
50              \
51 # new directory for storing sources and .deb files
52              && tempDir="$(mktemp -d)" \
53              && chmod 777 "$tempDir" \
54 # (777 to ensure APT's "_apt" user can access it too)
55              \
```

```
# save list of currently-installed packages so build dependencies can be cleanly removed later
			&& savedAptMark="$(apt-mark showmanual)" \
			\
# build .deb files from upstream's source packages (which are verified by apt-get)
			&& apt-get update \
			&& apt-get build-dep -y $nginxPackages \
			&& ( \
				cd "$tempDir" \
				&& DEB_BUILD_OPTIONS="nocheck parallel=$(nproc)" \
					apt-get source --compile $nginxPackages \
			) \
# we don't remove APT lists here because they get re-downloaded and removed later
			\
# reset apt-mark's "manual" list so that "purge --auto-remove" will remove all build dependencies
# (which is done after we install the built packages so we don't have to redownload any overlapping dependencies)
			&& apt-mark showmanual | xargs apt-mark auto > /dev/null \
			&& { [ -z "$savedAptMark" ] || apt-mark manual $savedAptMark; } \
			\
# create a temporary local APT repo to install from (so that dependency resolution can be handled by AP   as it should be)
			&& ls -lAFh "$tempDir" \
			&& ( cd "$tempDir" && dpkg-scanpackages . > Packages ) \
			&& grep '^Package: ' "$tempDir/Packages" \
			&& echo "deb [ trusted=yes ] file://$tempDir ./" > /etc/apt/sources.list.d/temp.list \
# work around the following APT issue by using "Acquire::GzipIndexes=false" (overriding "/etc/apt/ap conf.d/docker-gzip-indexes")
#	Could not open file /var/lib/apt/lists/partial/_tmp_tmp.ODWljpQfkE_._Packages - open (13: Pe ission denied)
#	...
#	E: Failed to fetch store:/var/lib/apt/lists/partial/_tmp_tmp.ODWljpQfkE_._Packages   Could not open file /var/lib/apt/lists/partial/_tmp_tmp.ODWljpQfkE_._Packages - open (13: Permission denied)
			&& apt-get -o Acquire::GzipIndexes=false update \
			;; \
	esac \
	\
	&& apt-get install --no-install-recommends --no-install-suggests -y \
					$nginxPackages \
					gettext-base \
					curl \
	&& apt-get remove --purge --auto-remove -y && rm -rf /var/lib/apt/lists/* /etc/apt/sources.list.d/ng x.list \
	\
# if we have leftovers from building, let's purge them (including extra, unnecessary build deps)
	&& if [ -n "$tempDir" ]; then \
		apt-get purge -y --auto-remove \
		&& rm -rf "$tempDir" /etc/apt/sources.list.d/temp.list; \
	fi \
# forward request and error logs to docker log collector
	&& ln -sf /dev/stdout /var/log/nginx/access.log \
	&& ln -sf /dev/stderr /var/log/nginx/error.log \
```

安装 Nginx 之后，继续配置镜像。

```
101 # create a docker-entrypoint.d directory
102         && mkdir /docker-entrypoint.d
103
104 COPY docker-entrypoint.sh /
105 COPY 10-listen-on-ipv6-by-default.sh /docker-entrypoint.d
106 COPY 20-envsubst-on-templates.sh /docker-entrypoint.d
107 COPY 30-tune-worker-processes.sh /docker-entrypoint.d
```

第 102 行创建工作目录/docker-entrypoint.d。

第 104~第 107 行的 COPY 指令，复制 4 个脚本文件到镜像中。如果目标路径不存在，会自动创建。COPY 与 ADD 指令功能类似，不过 ADD 支持压缩文件：先上传压缩文件，再解压缩。

```
108 ENTRYPOINT ["/docker-entrypoint.sh"]
109
```

第 108 行指令设置镜像的默认入口命令。入口命令会在启动容器时作为根命令执行，所有传入值作为该命令的参数，支持两种格式：

- ENTRYPOINT["executable"，"param1"，"param2"]，将通过 exec 调用执行。
- ENTRYPOINT command param1 param2，将通过 Shell 执行。

此外，CMD 指令指定的值将作为根命令的参数。

每个 Dockerfile 中只能有一个 ENTRYPOINT，当指定多个时，只有最后一个起效。在运行时，可以被--entrypoint 参数覆盖掉，如 docker run --entrypoint。

```
110 EXPOSE 80
111
```

第 110 行指令声明镜像内服务监听的端口。它的格式为：

```
EXPOSE<port>[<port>/<protocol>...]
```

例如：EXPOSE 22 80 8443。

该指令只是起到声明作用，并不会自动完成端口映射。如果要映射端口，在启动容器时可以使用-P 参数(Docker 主机会自动分配一个宿主机的临时端口)或-p HOST_PORT：CONTAINER_PORT 参数(具体指定所映射的本地端口)。

```
112 STOPSIGNAL SIGQUIT
113
```

第 112 行指令设置启动的容器接收退出的信号值。

```
114 CMD ["nginx", "-g", "daemon off;"]
```

第 114 行指令用来指定启动容器时默认执行的命令。

12.2 Dockerfile 的配置指令

Dockerfile 的配置指令用于配置镜像信息，如表 12-1 所示。

表 12-1　Dockerfile 的配置指令

指令	说明
ARG	定义构建镜像过程中使用的变量
FROM	指定基础镜像
LABEL	设置新镜像的元数据标签信息
EXPOSE	声明镜像内服务监听的端口
ENV	指定环境变量
ENTRYPOINT	指定镜像的默认入口命令
VOLUME	创建一个数据卷挂载点
USER	指定运行容器时的用户名或 UID
WORKDIR	配置工作目录
ONBUILD	创建子镜像时指定自动执行的操作指令
STOPSIGNAL	指定退出的信号值
HEALTHCHECK	指定健康检查方式
SHELL	指定默认的 Shell 类型

(1) ARG 指令定义创建镜像过程中使用的变量。

格式：ARG <name>[=<default value>]

Dockerfile 可以包括一个或多个 ARG 指令。在使用 build 子命令构建镜像时，可以通过 --build-arg list 来为变量赋值。当镜像构建成功后，ARG 指定的变量将不再存在(ENV 指定的变量将在镜像中保留)。

Docker 有一些内置的创建镜像时的变量，用户无须声明即可使用，包括 HTTP_PROXY、HTTPS_PROXY、FTP_PROXY、NO_PROXY(不区分大小写)。

```
1 $ cat Dockerfile
    FROM busybox
    ARG user1
    ARG buildno=1
    RUN echo $user1 > 1.txt

2 $ docker image build --build-arg user1=tom -t mytest1 .

3 $ docker run --rm -it mytest1 cat /1.txt
    tom
```

第 1 行命令显示 Dockerfile 文件，其中变量 user1 没有默认值，变量 buildno 的默认值为 1。

如果用户指定了 Dockerfile 中未定义的生成参数，则生成将输出警告。

(2) FROM 指令用于指定基础镜像。

格式：FROM [--platform=<platform>] <image> [AS <name>]

任何 Dockerfile 都需要有 FROM 指令，除了 ARG 指令外，它必须是第一条。如果是一个全新的镜像，需要指定基础镜像 scratch。还可以通过 TAG 指定一个精确的版本。

如果想通过一个 Dockerfile 创建多个镜像(如多步骤创建)，可以使用多个 FROM 指令(每个镜像指定一次)。

```
FROM scratch
ADD alpine-minirootfs-20220715-x86_64.tar.gz /
CMD ["/bin/sh"]
```

大多数 Linux 发行版本的官方镜像都是 FROM scratch，而其他服务类型的容器(HTTPD、Nginx、MySQL)，多数是基于某种 Linux 发行版本的容器。为了保证镜像精简，建议用体积较小的镜像(如 Alpine、Debian)作为基础镜像。

(3) LABEL 指令为镜像添加元数据标签信息。

格式：LABEL <key>=<value><key>=<value><key>=<value>...

这些信息可以用来辅助过滤出特定镜像，示例如下：

```
LABEL version="1.0.0-rc3"
LABEL author="test@github" date="2022-12-28"
LABEL description="This text illustrates \that label-values can span multiple lines."
```

(4) EXPOSE 指令声明镜像内服务监听的端口。

格式：EXPOSE <port>[<port>/<protocol>...]

该指令仅起到声明的作用，并不会自动完成端口映射，示例如下：

```
EXPOSE 22 80 8443
```

如果需要映射端口，在启动容器时可以使用-P 选项(Docker 主机会自动分配一个宿主机的临时端口)或-p HOST_PORT：CONTAINER_PORT 选项(具体指定所映射的本地端口)。

(5) ENV 指令用于设置环境变量。

格式：

ENV <key><value>

ENV <key>=<value>...

这些变量既可以被后续 RUN 指令引用，也可以由此镜像启动的容器使用。示例如下：

```
ENV APP_VERSION=1.0.0
ENV APP_HOME=/usr/local/app
ENV PATH $PATH:/usr/local/bin
```

这些环境变量在运行时可以被--env 选项的值覆盖掉。

(6) ENTRYPOINT 指令设置启动容器时自动执行的程序。

格式：

EXEC 格式：ENTRYPOINT ["executable", "param1", "param2"]

这是首选格式。

SHELL 格式：ENTRYPOINT command param1 param2

这将在 Shell 中执行。

如果希望容器启动后，立即执行一个应用程序或脚本，设置 ENTRYPOINT 是一种解决方法。示例如下：

```
1 $ mkdir test && cd test

2 $ vi Dockerfile
    FROM alpine
    ENTRYPOINT ["top"]

3 $ docker image build -t test1 .
    Sending build context to Docker daemon    2.048kB
    Step 1/2 : FROM alpine
     ---> d7d3d98c851f
    Step 2/2 : ENTRYPOINT ["top"]
     ---> Running in df442ad3fee3
    Removing intermediate container df442ad3fee3
     ---> 684dc21ddba3
    Successfully built 684dc21ddba3
    Successfully tagged test1:latest
```

第 3 行命令使用-t 选项指定镜像的名称，命令最后的句号表示当前目录。

```
4 $ docker history test1
    IMAGE           CREATED          CREATED BY                                        SIZE    COMMENT
    684dc21ddba3    12 seconds ago   /bin/sh -c #(nop) ENTRYPOINT ["top"]              0B
    d7d3d98c851f    12 days ago      /bin/sh -c #(nop) CMD ["/bin/sh"]                 0B
    <missing>       12 days ago      /bin/sh -c #(nop) ADD file:a2648378045615c37…  5.53MB

5 $ docker image inspect test1 | more

      "Cmd": [
          "/bin/sh",
          "-c",
          "#(nop) ",
          "ENTRYPOINT [\"top\"]"
      ],
      "Image": "sha256:d7d3d98c851ff3a95dbcb70ce09d186c9aaf7e25d48d55c0f99aae360aecfd53",
      "Volumes": null,
      "WorkingDir": "",
      "Entrypoint": [
          "top"
      ],

6 $ docker run --rm -it --name t1 test1
    Mem: 809696K used, 7028744K free, 9220K shrd, 3748K buff, 343260K cached
    CPU:   60% usr   40% sys    0% nic    0% idle    0% io    0% irq    0% sirq
    Load average: 0.00 0.00 0.00 3/205 6
      PID  PPID  USER     STAT   VSZ  %VSZ  CPU %CPU  COMMAND
        1     0  root     R     1604    0%    0   0%   top
```

从第 6 行命令的输出可以看出：启动容器后会自动执行 top 命令。

再打开一个新会话，查看容器中的进程信息，示例命令如下：

```
$ docker container exec -it t1 ps aux
  PID    USER         TIME   COMMAND
  1      root         0:00   top
  12     root         0:00   ps aux
```

从输出可以看出：容器中 top 命令的 PID 是 1，说明它是根进程。

(7) VOLUME 指令用于创建与容器生命周期一致的数据卷。

格式：

VOLUME ["/data"]

VOLUME /data

通过 VOLUME 指令创建的数据卷，适用于对存储 IO 要求比较高的场景。这种卷的生命周期与容器一致：创建容器时自动创建卷，删除容器时也会自动删除卷。

MongoDB 是一种开源的非关系数据库，它是存储的 IO 要求比较高的典型应用。官方版的 MongoDB 的 Dockerfile 如下：

```
……略……
VOLUME /data/db /data/configdb

ENV HOME /data/db

COPY docker-entrypoint.sh /usr/local/bin/
ENTRYPOINT ["docker-entrypoint.sh"]

EXPOSE 27017
CMD ["mongod"]
```

运行这个容器的时候，会在宿主机上自动创建两个卷，它们属于匿名卷，会被挂载到容器中的/data/db 和/data/configdb 这两个目录中。删除这个容器的时候，会自动删除这两个卷。

(8) USER 指令指定运行容器时的用户名(UID)和组名(GID)。

格式：

USER <user>[:<group>]

USER <UID>[:<GID>]

默认情况下，构建容器、在容器中运行命令，是以 root 身份进行的。为了降低安全风险，可以使用 USER 指令，在 Docker 容器内更改或切换到其他用户。

USER 指令用来设置用户名(或 UID)和可选的用户组(或 GID)，它们将用作随后的默认用户和组来执行 RUN 指令，并在运行时运行相关的 ENTRYPOINT 和 CMD 命令。

在使用 USER 指令之前，需要在容器内创建用户和组，示例如下：

```
FROM ubuntu
RUN apt-get -y update
RUN groupadd -r dbuser && useradd -r -g dbuser dbuser
USER dbuser
```

(9) WORKDIR 指令设置工作目录。

格式：WORKDIR /path/to/workdir

WORKDIR 指令为 RUN、CMD、ENTRYPOINT、COPY 和 ADD 指令设置工作目录。如果指定的目录不存在，会自动创建，即使它在任何后续 Dockerfile 指令中没有使用。

在 Dockerfile 中可以有多个 WORKDIR 指令。如果指定的是相对路径，则是与上一条 WORKDIR 指令相对的路径，示例如下：

```
……
WORKDIR /a
WORKDIR b
WORKDIR c
RUN pwd
```

示例中有 3 个 WORKDIR 指令，第 2、第 3 个指令使用的是相对路径，所以最终路径为/a/b/c。

(10) ONBUILD 指令设置创建“子”镜像时自动执行的操作指令。

格式：ONBUILD <INSTRUCTION>

ONBUILD 字面的意思是“在构建时”，但是 ONBUILD 指令并不是构建当前的镜像时进行的操作，而是指定在创建它的“子”镜像的时候，自动执行的操作指令，它对当前镜像是没有用的。

ONBUILD 指令是在“父”镜像中设置的，是当创建子镜像时自动执行的操作指令。如果基于“子”镜像再构建镜像，也就是“孙”镜像，是不会执行这些操作指令的。

ONBUILD 指令的作用类似于“触发器”，当使用这个镜像再制作另一个镜像的时候，这些指令就会被执行。触发器将在子镜像构建的上下文中执行。当为应用程序构建环境时，或者使用特定于用户的配置进行自定义的守护进程时，这个选项非常有用。示例如下：

```
# Dockerfile for ParentImage
[...]
ONBUILD ADD . /app/src
ONBUILD RUN /usr/local/bin/python-build --dir /app/src
[...]
```

这个 Dockerfile 有两个 ONBUILD 指令，一个使用 ADD 指令复制原代码文件，另一个使用 RUN 指令构建 python 应用。在构建父镜像时，不会执行操作指令，仅仅是设置一个触发器。

但是，当使用 docker build 命令创建子镜像 ChildImage 时，看到有 FROM ParentImage，触发器就会被触发，就会先执行 ParentImage 中配置的 ONBUILD 指令：

```
# Dockerfile for ChildImage
FROM ParentImage
```

这就相当于在 ChildImage 的 Dockerfile 中添加了如下指令：

```
#Automatically run the following when building ChildImage
ADD . /app/src
RUN /usr/local/bin/python-build --dir /app/src...
```

从这个示例可以看出，ONBUILD 指令在创建专门用于自动编译、检查等操作的基础镜像时十分有用。

(11) STOPSIGNAL 指令设置容器接收退出的信号值。

格式：STOPSIGNAL signal

容器的进程收到信号可以进行相关的操作。信号的格式既可以是 SIG 开头的信号名(例如 SIGKILL)，也可以是匹配内核系统调用表中的无符号数字(例如 9)。

如果Dockerfile中没有设置退出的信号，那么将发送默认的SIGTERM。还可以在使用docker run 和 docker create 的时候，使用--stopsignal 选项，来覆盖这个指令。

(12) HEALTHCHECK 指令设置健康检查方式。

格式：

HEALTHCHECK [OPTIONS] CMD command

HEALTHCHECK NONE

选项：

- --interval=DURATION：默认为 30s。
- --timeout=DURATION：默认为 30s。
- --start-period=DURATION：默认为 0s。
- --retries=N：默认为 3 次。

HEALTHCHECK 指令配置所启动容器如何进行健康检查。

判断 Web 服务器的容器状态，示例如下：

```
……略……
HEALTHCHECK --interval=5m --timeout=3s \
  CMD curl -f http://localhost/ || exit 1
……略……
```

每 5 分钟左右检查一次，以确保 Web 服务能够在 3 秒内提供服务。

(13) SHELL 指令设置默认的 Shell 类型。

格式：SHELL ["executable", "parameters"]

如果未设置，默认值为：

- Linux ["/bin/sh"，"-c"]
- Windows ["cmd"，"/S"，"/C"]

12.3 Dockerfile 的操作指令

与配置指令相比，Dockerfile 的操作指令比较少。操作指令用于构建时执行的具体操作，如表 12-2 所示。

表 12-2　Dockerfile 的操作指令

指令	说明
ADD	添加内容到镜像
COPY	复制内容到镜像
CMD	启动容器时指定默认执行的命令
RUN	运行指定命令

(1) ADD 指令添加文件到镜像。

格式：

ADD [--chown=<user>:<group>] <src>... <dest>

ADD [--chown=<user>:<group>] ["<src>",... "<dest>"]

<src>可以是：

- Dockerfile 所在目录的一个相对路径(文件或目录)。
- 一个 URL。
- 一个 tar 文件(自动解压为目录)。

<dest>可以是：

- 镜像内绝对路径。
- 相对于工作目录(WORKDIR)的相对路径。

所有新文件和目录都是使用 UID 和 GID 0 创建的，除非使用--chown 标志指定了给定的用户名、组名或 UID/GID 组合来修改所有权。示例如下：

```
ADD --chown=55:mygroup files* /somedir/
ADD --chown=bin files* /somedir/
ADD http://foo.com/bar.go /somedir/somefile.go
```

(2) COPY 指令复制内容到镜像。

格式：

COPY [--chown=<user>:<group>] <src>... <dest>

COPY [--chown=<user>:<group>] ["<src>",... "<dest>"]

与 ADD 指令相比，COPY 指令操作简单，可以把本地的文件复制到容器镜像中。

<src>为 Dockerfile 所在目录的相对路径下的文件或目录，如果目标路径<dest>不存在，则会自动创建。如果<dest>不是以/结尾，则将被视为常规文件。示例如下：

```
COPY hom* /mydir/
```

添加以 hom 开头的所有文件。

```
COPY hom?.txt /mydir/
```

将“？”替换为任何单个字符，例如“home.txt”。

```
COPY test.txt relativeDir/
```

<dest>是一个绝对路径，或相对于 WORKDIR 的路径。使用相对路径，并将 test.txt 添加到<WORKDIR>/relativeDir/：

```
COPY test.txt /absoluteDir/
```

使用绝对路径，并将 test.txt 添加到/absoluteDir/：

```
COPY test.txt /file
```

如果<dest>未以尾部斜杠结尾，则将被视为常规文件，<src>的内容将被写入<dest>。

```
COPY arr[[]0].txt /mydir/
```

复制包含特殊字符(如[和])的文件或目录时，需要按照 Go 语言规则转义这些路径，以防止它们被视为匹配模式。例如，复制名为 arr[0].txt 的命令如下：

```
COPY --chown=55:mygroup files* /somedir/
```

所有新文件和目录默认都是使用 UID 和 GID 0 创建的，可以通过--chown 标志来修改所有权。

(3) CMD 指令指定启动容器时默认执行的命令。

格式：

CMD ["executable","param1","param2"]

CMD ["param1","param2"]

CMD command param1 param2

CMD 指令与 ENTRYPOINT 指令类似，都是指定容器启动后的执行命令。如果有多个，都是最后一个生效。

CMD 指令是容器运行默认参数，很容易被 docker run 指定的命令覆盖。如果 CMD 与 ENTRYPOINT 同时存在，CMD 指令将作为 ENTRYPOINT 指令的默认参数。

ENTRYPOINT 指令是容器的主执行程序，只能被--entrypoint 指定的命令覆盖。

在 HTTPD 的 Dockerfile 中没有 ENTRYPOINT 指令，只有 CMD 指令，执行 httpd-foreground 脚本用于自动启动 httpd 进程，示例命令如下：

```
$ docker run --rm -d --name test1 -P httpd

$ docker ps
CONTAINER ID IMAGE      COMMAND             CREATED         STATUS
847b230e3361  httpd     "httpd-foreground"  9 seconds ago   Up 8 seconds
 PORTS                                      NAMES
0.0.0.0:49153->80/tcp, :::49153->80/tcp     test1

$ curl localhost:49153
<html><body><h1>It works!</h1></body></html>

$ docker stop test1
```

CMD 指令很方便被替换，示例命令如下：

```
$ docker run --rm -it --name test1 -P httpd sh

$ docker ps -a
CONTAINER ID IMAGE    COMMAND CREATED          STATUS
f20c3cdb2407  httpd   "sh"    50 seconds ago   Up 49 seconds
 PORTS                                    NAMES
0.0.0.0:49154->80/tcp, :::49154->80/tcp   test1

$ curl localhost:49175
curl: (7) Failed to connect to localhost port 49175: Connection refused
```

(4) RUN 指令运行指定的指令。

格式：

RUN <command>

RUN ["executable", "param1", "param2"]

每条 RUN 指令将在当前镜像基础上执行命令，并提交为新的镜像层。建议通过合并多个 RUN 指令来减少容器的层数。当命令较长时，可以使用\进行换行，示例如下：

```
RUN apt-get update \
    && apt-get install -y libsnappy-dev zlib1g-dev libbz2-dev \
    && rm -rf /var/cache/apt \
    && rm -rf /var/lib/apt/lists/*
```

12.4 创建镜像

撰写完 Dockerfile 之后，就可以通过 build 子命令来创建镜像，它的语法格式如下：

```
docker build [OPTIONS] PATH|URL|-
```

build 子命令将读取指定路径下的 Dockerfile，并将该路径下所有文件作为上下文(context)发送给 Docker 服务端。在构建过程中，如果上下文过大，会导致发送大量数据给服务端，延缓创建过程。因此除非是生成镜像所必需的文件，就不要放到上下文路径下。

如果希望使用非上下文路径下的 Dockerfile，可以通过-f 选项来指定路径。

Docker 服务端在校验 Dockerfile 格式之后，会逐条执行其中定义的指令，遇到 ADD、COPY 和 RUN 等指令就生成一层新的镜像。如果最后成功创建镜像，会输出最终镜像的 ID。

通常需要使用-t 选项指定生成镜像的标签信息。重复使用多次-t 选项，为镜像添加多个标签。例如，希望生成镜像标签为 builder/first_image:1.0.0，上下文路径为/tmp/docker_builder/，则执行下面的构建命令：

```
$ docker build -t builder/first_image:1.0.0 /tmp/docker_builder/
```

12.4.1 命令选项

build 子命令通过一系列的选项来调整创建镜像过程的行为。主要选项如下：

- --add-host：添加自定义的主机到 IP 映射。
- --build-arg：设置构建时变量。
- --cache-from：当作缓存源的镜像。
- --cgroup-parent：容器的可选父 cgroup。
- --compress：使用 gzip 压缩构建上下文。
- --cpu-period：限制 CPU 的 CFS(completely fair scheduler，完全公平调度器)周期。
- --cpu-quota：限制 CPU 的 CFS 配额。
- -c, --cpu-shares：CPU 份额(相对权重)。
- --cpuset-cpus：允许执行的 CPU(0-3,0,1)。
- --cpuset-mems：允许执行的 MEMs(0-3,0,1)。

- --disable-content-trust：跳过镜像验证(默认为 true)。
- -f, --file：Dockerfile 的名称(默认值为 PATH/Dockerfile)。
- --force-rm：删除中间过程的容器。
- --iidfile：将镜像 ID 写入文件。
- --isolation：容器隔离技术。
- --label：设置镜像的元数据。
- -m, --memory：内存限制。
- --memory-swap：内存加交换空间的整体限制。
- --network：在构建期间设置 RUN 指令的网络模式。
- --no-cache：构建镜像时不使用缓存。
- --pull：始终下拉镜像的更新版本。
- -q, --quiet：成功时抑制生成输出和打印镜像 ID。
- --rm：成功生成后删除中间容器(默认为 true)。
- --security-opt：安全选项。
- --shm-size：/dev/shm 的大小。
- -t, --tag：名称和可选的“name:tag”格式的标记。
- --target：将目标构建阶段设置为构建。
- --ulimit：Ulimit 选项(默认值[])。

12.4.2 父镜像的选择

大部分情况下，生成新的镜像都需要通过 FROM 指令来指定父镜像。父镜像是生成镜像的基础，会直接影响到所生成镜像的大小和功能。可以选择两种镜像作为父镜像，一种是所谓的基础镜像(base image)，另外一种是普通的镜像(往往由第三方创建，是基于基础镜像的镜像)，如图 12-4 所示。

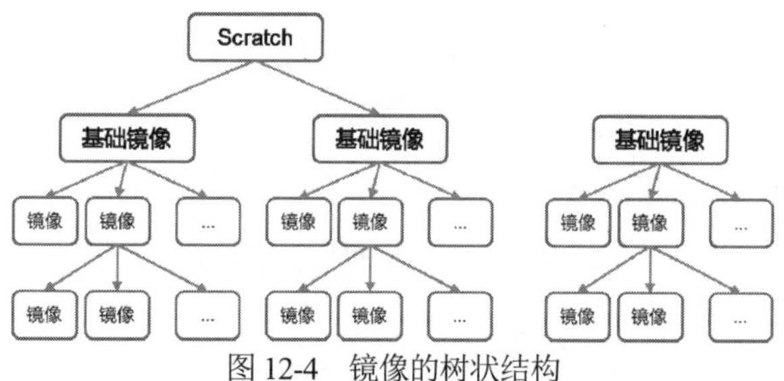

图 12-4　镜像的树状结构

基础镜像比较特殊，其 Dockerfile 中往往不存在 FROM 指令，或者基于 scratch 镜像(FROM scratch)，这意味着其在整个镜像树中处于根的位置，示例如下：

```
FROM scratch
ADD binary /
CMD ["/binary"]
```

这个 Dockerfile 定义了一个简单的基础镜像，将编译好的二进制可执行文件 binary 复制到镜像中，运行容器时执行 binary 命令。

普通镜像也可以作为父镜像来使用，包括常见的 Busybox、Debian、Ubuntu 等。

12.4.3 使用.dockerignore 文件

可以通过.dockerignore 文件(每一行添加一条匹配模式)来让 Docker 忽略匹配路径或名称的文件，在创建镜像时不将无关数据发送到服务端。

在下面的例子中包括了 3 行忽略的模式(第一行为注释)：

```
$ cat .dockerignore
  # comment
  */temp*
  */*/temp*
  temp?
```

配置文件中模式语法支持 Go 语言风格的路径正则格式：

- “*”表示任意多个字符。
- “?”代表单个字符。
- “!”表示不匹配(即不忽略指定的路径或文件)。

12.4.4 多步骤创建

有一些需要编译的应用(例如 C、Go 或 Java 语言)，通常不会将源代码打包到正式生产用的容器中。在 Docker 的早期版本(17.05 版本之前)中，需要准备两个环境的 Docker 镜像：

- 编译环境镜像：包括完整的编译引擎、依赖库等，用于将源代码编译应用为二进制文件。这个镜像往往比较庞大。
- 运行环境镜像：利用编译好的二进制文件，运行应用对外提供服务，这是最终应用的镜像。由于不需要编译环境，所以体积比较小。

现在可以使用单一的 Dockerfile 文件，通过多步骤创建来简化构建过程。最终生成的运行环境镜像不包括源代码的、精简的镜像。

示例：有一个 Go 语言编写的应用，文件名是 main.go，执行的结果是输出“Hello，Docker”字符串，代码如下：

```
$ cat -n main.go
  // main.go will output "Hello, Docker"
  package main
  import (
        "fmt"
  )
  func main() {
        fmt.Println("Hello, Docker")
  }
```

构建用的 Dockerfile 文件如下：

```
$ cat -n Dockerfile
  # define stage name as builder
  FROM golang:1.9 AS builder
  RUN mkdir -p /go/src/test
  WORKDIR /go/src/test
```

```
5 COPY main.go .
6 RUN CGO_ENABLED=0 GOOS=linux go build -o app .
7
8 FROM alpine:latest
9 WORKDIR /root
10 # copy file from the builder stage
11 COPY --from=builder /go/src/test/app .
12 CMD ["./app"]
```

构建过程分为以下两个步骤：

(1) 使用 golang:1.9 镜像编译应用二进制文件 app。

(2) 使用精简的镜像 alpine:latest 作为运行环境。

第 2 行指令，通过 AS 关键字给这个步骤起一个别名，叫 builder。在第 11 行指令中，使用--from=builder 来引用它。当然也可以不用别名而是使用编号，由于第 1 个步骤编号为 0，所以第 11 行指令可以修改为--from=0。

第 3 行指令在容器中创建一个目录。

第 4 行指令切换到这个目录下，把它当作工作目录。

第 5 行指令将源代码文件复制到这个目录中。

第 6 行指令调用编译命令。前面设置两个环境变量。编译出的文件就在当前目录下，文件名是 app。

第 8 行指令指定正式运行用的镜像。

第 9 行指令创建工作目录，并切换到这个目录中。

第 11 行指令从名为 builder 步骤生成的镜像中复制文件(/go/src/test/app)到当前目录下。

第 12 行指令设置自动执行的命令。

12.5 习题

1. [单选题]不属于 Docker 创建镜像的方法是(　　)。

   A. 基于 Dockerfile 创建　　B. 基于 Makefile 创建

   C. 基于现有镜像创建　　D. 基于本地模板创建

2. [单选题]下列关于 Dockerfile 的描述，错误的是(　　)。

   A. Dockerfile 是由一组指令组成的文件

   B. Docker 程序读取 Dockerfile 中的指令生成指定的镜像

   C. Dockerfile 每行支持一条指令，每条指令最多可携带一个参数

   D. Dockerfile 由镜像信息、维护者、操作指令和容器启动执行的指令组成

3. [单选题]Dockerfile 中构建镜像的命令是(　　)。

   A. docker build　　B. docker commit

   C. docker export　　D. docker create

4. [单选题]Dockerfile 中指定镜像源的参数是(　　)。

A. From　　B. Run

C. Search　　D. Commit

5. [单选题]使用 docker build 构建镜像的时候，通过(　　)指定特定的 Dockerfile 文件。

A. -c　　B. -d

C. -f　　D. -e

第 13 章

Docker实战案例

“纸上得来终觉浅，绝知此事要躬行”，本章将通过几个经典案例来深入学习如何在生产实践中使用容器技术。

本章要点：

◎ Linux 操作系统镜像

◎ 为镜像添加 SSH 服务

◎ Web 服务

◎ 数据库服务

13.1 Linux 操作系统镜像

在多个 Linux 发行版中，其中一些版本以软件包丰富而出名，一些版本用户体验好、适合用于桌面操作系统，一些版本则运行稳定、适合当作服务器操作系统。在容器中采用何种发行版本，取决于具体需求，如图 13-1 所示。

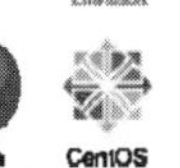

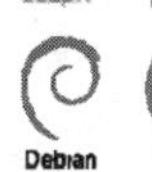

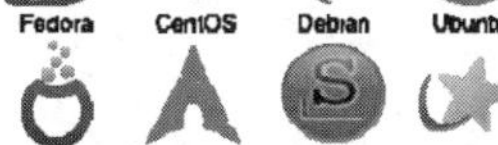

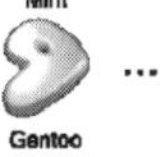

图 13-1　Linux 发行版本的组成

13.1.1 BusyBox

BusyBox 是一个集成了数百个最常用 Linux 命令(如 cat、echo、grep、mount、telnet 等)的精简工具箱，它的大小不到 2MB，被誉为"Linux 系统的瑞士军刀"。BusyBox 既可在多款 POSIX 系列的操作系统中直接执行，如 Linux(包括 Android)、Hurd、FreeBSD 等，也可以当作容器来执行。

```
$ docker images busybox
  REPOSITORY        TAG       IMAGE ID        CREATED         SIZE
  busybox           latest    62aedd01bd85    2 months ago    1.24MB
```

Docker Hub 提供的 BusyBox 的 Dockerfile 如下：

```
FROM scratch
ADD busybox.tar.xz /
CMD ["sh"]
```

BusyBox 镜像虽然小，但包括了大量常见的 Linux 命令(目前有 400 多个)。

```
$ docker run --rm -it busybox sh
/ # ls -l /bin | wc -l
403
```

除了进入容器之外，还可以直接使用其中的命令：

```
$ docker run --rm -it busybox nslookup www.baidu.com
Server:    8.8.8.8
Address:   8.8.8.8:53

Non-authoritative answer:
www.baidu.com      canonical name = www.a.shifen.com
Name:      www.a.shifen.com
Address:   110.242.68.3
Name:      www.a.shifen.com
Address:   110.242.68.4

*** Can't find www.baidu.com: No answer
```

因为 BusyBox 容器小巧，所以很适合作为基本镜像。下面演示如何通过它构建一个自定义的 Web 服务器。

```
1 $ mkdir build && cd build

2 $ vi server.go
    package main

    import (
            "fmt"
            "net/http"
    )

    func handler(w http.ResponseWriter, r *http.Request) {
            fmt.Fprintf(w, "Hello World!")
```

```
	}

	func main() {
		http.HandleFunc("/", handler)
		fmt.Println("Server running...")
		http.ListenAndServe(":8080", nil)
	}
```

接下来通过 Go 语言编写一个简单的 Web 服务器，即通过多步骤创建来制作镜像。

```
3 $ vi Dockerfile
  FROM golang AS builder
  WORKDIR /src
  COPY server.go .
  RUN CGO_ENABLED=0 GOOS=linux go build server.go

  FROM busybox
  EXPOSE 8080
  WORKDIR /
  COPY --from=builder /src/server .
  CMD ["/server"]


4 $ docker build -t go-server .

5 $ docker image ls | grep go-server
  go-server     latest     6693b782cd2a    2 minutes ago    7.69MB
```

构建出的镜像也很小巧，只有 7.7MB。

```
6 $ docker run --rm -d --name web1 -P go-server

7 $ docker ps
  CONTAINER ID  IMAGE     COMMAND     CREATED         STATUS
  4abf03ca1a18  go-server "/server"   7 seconds ago   Up 6 seconds
  PORTS                                          NAMES
  0.0.0.0:49153->8080/tcp, :::49153->8080/tcp    web1

8 $ curl localhost:49153
Hello World!
```

13.1.2 Alpine

Alpine 的 Docker 官方镜像非常小巧，最新的版本只有 5MB 多点。包括 Docker 官方在内的很多组织都推荐使用 Alpine 作为基础镜像，它具有很多优势：镜像下载速度快、安全性高、主机之间的切换方便、占用磁盘空间少。

Alpine 比 BusyBox 功能更完善，还可以通过包管理工具 apk 进行软件包的管理。以安装 curl 命令为例，命令如下：

```
$ docker run --rm -it alpine sh
```

```
/ # curl
sh: curl: not found

/ # apk update

/ # apk add curl
(1/5) Installing ca-certificates (20220614-r3)
(2/5) Installing brotli-libs (1.0.9-r9)
(3/5) Installing nghttp2-libs (1.51.0-r0)
(4/5) Installing libcurl (7.87.0-r0)
(5/5) Installing curl (7.87.0-r0)
Executing busybox-1.35.0-r29.trigger
Executing ca-certificates-20220614-r3.trigger
OK: 9 MiB in 20 packages

/ # curl baidu.com
<html>
<meta http-equiv="refresh" content="0;url=http://www.baidu.com/">
</html>
```

13.1.3 Debian/Ubuntu

Debian 和 Ubuntu 都是目前流行的 Debian 家族的服务器操作系统。Docker Hub 上提供了它们的官方镜像，国内外云服务商都提供了完整虚拟机及容器的支持。Debian/Ubuntu 的软件仓库丰富、参考资料完善，如果使用中遇到问题，基本都可以找到解决方法。

如果官方镜像中没有需要的软件(例如 CURL)，可以通过软件仓库来进行安装。既可以在容器中添加，也可以通过 Dockerfile 在新镜像中添加，例如：

```
1 $ docker run --rm -it ubuntu curl baidu.com
    docker: Error response from daemon: failed to create shim task: OCI runtime create failed: runc create failed: unable to start container process: exec: "curl": executable file not found in $PATH: unknown.
    ERRO[0000] error waiting for container: context canceled

2 $ mkdir build && cd build

3 $ vi Dockerfile
    FROM ubuntu
    RUN apt-get update && apt-get install -y curl \
        && apt-get -y clean && rm -rf /var/lib/apt/lists/
    CMD ["bash"]

4 $ docker build -t myubuntu .

5 $ docker images | grep ubuntu
    myubuntu      latest    8f50ba0c57bd    17 seconds ago      84.7MB
    ubuntu        latest    a8780b506fa4    7 weeks ago         77.8MB

6 $ docker run --rm -it myubuntu curl baidu.com
    <html>
    <meta http-equiv="refresh" content="0;url=http://www.baidu.com/">
    </html>
```

13.1.4 CentOS/Fedora

CentOS 和 Fedora 也是流利的 Linux 发行版本，有着丰富的软件仓库和参考资料。不过充当容器，就不如以前几个发行版本了。

由于 CentOS 8 已经停止更新，所以不能直接通过在线软件仓库进行软件安装。在安装软件包之前，需要对镜像中的软件仓库的 URL 进行修改。例如，要安装 bind-utils 软件包，示例命令如下：

```
1 $ docker run --rm -it centos cat /etc/system-release
    CentOS Linux release 8.4.2105

2 $ mkdir build && cd build

3 $ vi Dockerfile
    FROM centos
    RUN sed -i -e 's/mirrorlist/#mirrorlist/g' \
    -e 's|#baseurl=http://mirror.centos.org|baseurl=https://vault.centos.org|g' \
      /etc/yum.repos.d/*.repo \
      && yum -y update \
      && yum -y install bind-utils \
      && yum -y clean all
    CMD ["bash"]

4 $ docker build -t mycentos .

5 $ docker images | grep centos
    mycentos      latest     a42f819ae247     5 minutes ago        572MB
    centos        latest     5d0da3dc9764     15 months ago        231MB

6 $ docker run --rm -it mycentos nslookup baidu.com
    Server:          8.8.8.8
    Address:         8.8.8.8#53

    Non-authoritative answer:
    Name:    baidu.com
    Address: 110.242.68.66
    Name:    baidu.com
    Address: 39.156.66.10
```

13.2 为镜像添加 SSH 服务

很多 Linux 系统管理员习惯通过 SSH 服务来远程登录服务器，但是 Docker 的 Linux 镜像大多数没有 SSH 服务(推荐使用 attach、exec 子命令进行连接)。如果必须使用 SSH，还可以在镜像中添加 SSH 服务。

下面将以 Ubuntu 镜像为例，介绍如何添加 SSH 组件。

```
1 $ mkdir build && cd build
```

首先创建一个新工作目录，可以避免将无用文件上传到镜像。

```
2 $ ssh-keygen -t rsa
    Generating public/private rsa key pair.
    Enter file in which to save the key (/home/tom/.ssh/id_rsa):
    /home/tom/.ssh/id_rsa already exists.
    Overwrite (y/n)? y
    Enter passphrase (empty for no passphrase):
    Enter same passphrase again:
    Your identification has been saved in /home/tom/.ssh/id_rsa
    Your public key has been saved in /home/tom/.ssh/id_rsa.pub
    The key fingerprint is:
    SHA256:4ohtMPfELutkJyGmJ6aPQH0mIYTGzx1q+Acu7epUIE0 tom@docker1
    The key's randomart image is:
    +---[RSA 3072]----+
    |o.E              |
    |o=     .         |
    |oo=.o .          |
    |..+*.o           |
    | .X++o+ S        |
    |.+.@+O .         |
    |+o= X =          |
    |=+ = =           |
    |++o.o            |
    +----[SHA256]-----+

3 $ ls -l ~/.ssh/
    total 16
    -rw-------. 1 tom tom    2590 Dec 24 18:23    id_rsa
    -rw-r--r--. 1 tom tom    565  Dec 24 18:23    id_rsa.pub
    -rw-------. 1 tom tom    1030 Aug 4 08:16     known_hosts

4 $ cp ~/.ssh/id_rsa.pub authorized_keys
```

在宿主机上生成 SSH 密钥对，并创建 authorized_keys 文件。

```
5 $ vi run.sh
    #!/bin/bash
    /usr/sbin/sshd -D

6 $ chmod +x run.sh
```

为镜像准备启动 SSH 服务的脚本 run.sh，并添加可执行权限。

```
7 $ vi Dockerfile
    # 设置基础镜像
    FROM ubuntu

    # 设置作者的信息
    MAINTAINER docker_user (user@docker.com)
```

```
RUN apt-get update

#安装 ssh 服务
RUN apt-get update && apt-get install -y openssh-server

RUN mkdir -p /var/run/sshd && mkdir -p /root/.ssh

#取消 pam 限制
RUN sed -ri 's/session    required    pam_loginuid.so/#session    required    pam_loginuid.so/g' /etc/pam.d/sshd

#复制配置文件到相应位置，并赋予脚本可执行权限
ADD authorized_keys /root/.ssh/
ADD run.sh /run.sh
RUN chmod 755 /run.sh

#开放端口
EXPOSE 22

#设置自启动命令
CMD ["/run.sh"]

8 $ ls -l
total 12
-rw-r--r--. 1 tom tom   565  Dec 24  18:25   authorized_keys
-rw-r--r--. 1 tom tom   593  Dec 24  18:27   Dockerfile
-rwxr-xr-x. 1 tom tom   31   Dec 24  18:26   run.sh

9 $ docker build -t sshd-ubuntu .
```

第 9 行命令创建镜像。

```
10 $ docker images | grep ubuntu
sshd-ubuntu    latest    e16ec9c3c941   About a minute ago   228MB
ubuntu         latest    a8780b506fa4   7 weeks ago          77.8MB

11 $ docker run --rm --name t1 -d -p 10022:22 sshd-ubuntu /run.sh

12 $ docker ps -a
    CONTAINER ID    IMAGE          COMMAND      CREATED           STATUS

    10a2b35c673d    sshd-ubuntu    "/run.sh"    39 seconds ago    Up 37 seconds
PORTS                                      NAMES
0.0.0.0:10022->22/tcp, :::10022->22/tcp    t1
```

通过 SSH 客户端进行测试连接。

```
$ ssh root@localhost -p 10022

$ ssh root@192.168.1.231 -p 10022
```

13.3 Web 服务

Web 服务和应用是目前互联网技术领域的常用技术。著名的互联网咨询服务公司 Netcraft 公司 2022 年的网络服务器调查报告显示：Web 服务器软件占用率的前两名分别是 Nginx 和 Apache HTTPD。

13.3.1 Nginx

Nginx 是一款功能强大的开源 Web 服务器、反向代理服务器，支持 HTTP、HTTPS、SMTP、POP3、IMAP 等协议。它可以作为 Web 服务器、HTTP 的缓存和负载均衡器。

Docker 官方提供的 Nginx 容器的默认配置参数可以满足大多数应用场景。在使用中，主要考虑如何将网站文件与 Nginx 容器解耦，最常用的解决方案是通过卷或绑定挂载。

下面通过绑定挂载将本地/www 目录挂载给 Nginx 容器，语句如下：

```
1 $ mkdir ~/www && cd www

2 $ echo "Test site" > index.html

3 $ docker run --rm --name web1 -d -P -v "$PWD":/usr/share/nginx/html nginx

4 $ docker ps -a
    CONTAINER ID  IMAGE   COMMAND                   CREATED         STATUS     x
    9a9b57459cef  nginx   "/docker-entrypoint.…"    12 seconds ago  Up 11 seconds
PORTS                                    NAMES
0.0.0.0:49153->80/tcp, :::49153->80/tcp  web1

5 $ curl localhost:49153
    Test site

6 $ echo v2 >> index.html

7 $ curl localhost:49153
    Test site
    v2
```

如果现有的 Nginx 配置无法满足要求，或者需要增加功能(例如启用 HTTPS)，则可以修改配置文件或创建自定义镜像，语句如下：

```
$ docker cp web1:/etc/nginx/nginx.conf .
$ docker cp web1:/etc/nginx/conf.d/default.conf .
$ vi nginx.conf
$ vi default.conf
……略……
```

13.3.2 Apache HTTPD

Apache HTTP 是一个高稳定性的企业级的开源 Web 服务器，由于其良好的跨平台和安全性，被广泛应用在多种平台和操作系统上。

官方 Apache HTTP 容器的配置与 Nginx 配置方法基本相同。对于小型的网站，也可以将网站页面直接打包到容器之中，示例如下：

```
1 $ mkdir src && cd src

2 $ mkdir public-html

3 $ echo "Test site" > public-html/index.html

4 $ vi Dockerfile
    FROM httpd
    COPY ./public-html/ /usr/local/apache2/htdocs

5 $ tree
    .
    ├── Dockerfile
    └── public-html
        └── index.hmtl

    1 directory, 2 files

6 $ docker build -t myhttpd .

7 $ docker images | grep httpd
    myhttpd      latest    764adb5254eb    14 seconds ago    145MB
    httpd        latest    8653efc8c72d    5 weeks ago       145MB

8 $ docker run --rm --name web1 -d -P myhttpd

9 $ docker ps -a
    CONTAINER ID   IMAGE     COMMAND               CREATED          STATUS
    54eea393f3c7   myhttpd   "httpd-foreground"   7 seconds ago    Up 7 seconds
PORTS                                       NAMES
0.0.0.0:49153->80/tcp, :::49153->80/tcp    web1

10 $ curl localhost:49153
Test site
```

13.4 数据库服务

数据库领域流行着一句话：没有一种数据库类型适合所有场景。在数据库选型时，要考虑的因素很多，包括数据格式、数据大小、查询频率、访问速度、数据保留期、可扩展性、总存储要求、持久性等。

可以简单地将数据库分为关系型(SQL)和非关系型(NoSQL)两种。

关系数据库是建立在关系模型基础上的数据库，借助于集合代数等数学概念和方法来处理数据库中的数据，支持复杂的事务处理和结构化查询。代表性的产品有 MySQL、Oracle、PostgreSQL、MariaDB、SQL Server 等。

非关系数据库是新兴的数据库技术，它放弃了传统关系型数据库的部分强致性限制，从而带来了性能上的提升，使其更适用于需要大规模并行处理的场景。代表性的产品有 MongoDB、Redis 等。

13.4.1 MySQL

MySQL 是全球最流行的开源关系型数据库之一，由于其具有高性能、成熟可靠、高适应性、易用性等特性而得到广泛应用。

在生产环境中，推荐使用官方镜像。使用官方 MySQL 镜像，需要注意几个环境变量。启动 MySQL 镜像时，可以通过在 docker run 命令行上传递一个或多个环境变量来调整 MySQL 实例的配置。

注意

如果使用已包含数据库的数据目录启动容器，则新设置的变量将不起作用。在容器启动时，任何预先存在的数据库都将始终保持不变。

当前版本使用的变量主要有：

- MYSQL_ROOT_PASSWORD：必需的参数，用于设置 root 账户的密码。
- MYSQL_DATABASE：启动时创建的数据库的名称。
- MYSQL_USER，MYSQL_PASSWORD：创建新用户和设置该用户的密码。
- MYSQL_ALLOW_EMPTY_PASSWORD：允许使用空密码的 root 用户启动容器。
- MYSQL_RANDOM_ROOT_PASSWORD：为 root 用户生成随机初始密码。
- MYSQL_ONETIME_PASSWORD：在第一次登录后，必须修改 root 的密码。
- MYSQL_INITDB_SKIP_TZINFO：禁用时加载。

提示

对变量的详细说明请参见 https://hub.docker.com/_/mysql。

MySQL 镜像既可以充当服务器端，也可以用于客户端，示例命令如下：

```
1 $ docker run -d --name dbsrv1 -e MYSQL_ROOT_PASSWORD=my-secret-pw mysql
2 $ docker run -it --rm mysql mysql -hsome.mysql.host -usome-mysql-user –p
```

第 1 行命令运行容器提供数据库服务，dbsrv1 是容器名称，my-secret-pw 为数据库的 root 用户密码。

第 2 行命令是使用客户端，连接到远程的 MySQL 实例(some.mysql.host)，用户名为 some-mysql-user。

为了提高存储 IO 性能，当前版本的 MySQL 镜像默认使用内部卷来存储/var/lib/mysql 中的数据。MySQL 镜像的 Dockerfile 如下：

```
……
VOLUME /var/lib/mysql

COPY docker-entrypoint.sh /usr/local/bin/
ENTRYPOINT ["docker-entrypoint.sh"]

EXPOSE 3306 33060
CMD ["mysqld"]
```

使用默认的内部卷有一些缺点：内部卷的生命周期与容器一致，同时也不方便进行管理。建议使用卷或绑定挂载来管理数据存储，示例命令如下：

```
1 $ docker run --name dbsrv1 -d -v ~/datadir:/var/lib/mysql \
-e MYSQL_ROOT_PASSWORD=secret-pw mysql

2 $ docker ps -a
    CONTAINER ID   IMAGE   COMMAND                  CREATED         STATUS
    5162e14e3f79   mysql   "docker-entrypoint.s…"   2 minutes ago   Up About a minute
PORTS                  NAMES
3306/tcp, 33060/tcp    dbsrv1

3 $ ls ~/datadir/ -sh
    total 98M
    4.0K  auto.cnf                8.2M '#ib_16384_1.dblwr'   12K   performance_schema
    2.9M  binlog.000001           8.0K  ib_buffer_pool       4.0K  private_key.pem
    4.0K  binlog.000002           12M   ibdata1              4.0K  public_key.pem
    4.0K  binlog.index            12M   ibtmp1               4.0K  server-cert.pem
    4.0K  ca-key.pem              4.0K '#innodb_redo'        4.0K  server-key.pem
    4.0K  ca.pem                  0 '#innodb_temp'           0 sys
    4.0K  client-cert.pem         0     mysql                16M   undo_001
    4.0K  client-key.pem          30M   mysql.ibd            16M   undo_002
    192K '#ib_16384_0.dblwr'      0     mysql.sock
```

还可以使用专用的“数据容器”来存储数据，示例命令如下：

```
1 $ docker create -v /var/lib/mysql --name data1 mysql

2 $ docker run --name dbsrv2 -d --volumes-from data1 \
-e MYSQL_ROOT_PASSWORD=secret-pw mysql

3 $ docker ps -a
   CONTAINER ID   IMAGE   COMMAND                  CREATED          STATUS         PORTS                 NAMES
   4daa0c6469e3   mysql   "docker-entrypoint.s…"   7 seconds ago    Up 6 seconds   3306/tcp, 33060/tcp   dbsrv2
   4d35262ea54f   mysql   "docker-entrypoint.s…"22 seconds ago   Created                              data1
```

13.4.2 MariaDB

MariaDB 数据库管理系统是 MySQL 的一个分支，完全兼容 MySQL，包括 API 和命令行，所以可以轻松地成为 MySQL 的替代品。

为了兼容 MySQL，MariaDB 容器的环境变量有两种风格：MARIADB 开头的和 MYSQL 开头的，它们是完全等效的。

MARIADB_ROOT_PASSWORD / MYSQL_ROOT_PASSWORD
MARIADB_ALLOW_EMPTY_ROOT_PASSWORD / MYSQL_ALLOW_EMPTY_PASSWORD
MARIADB_RANDOM_ROOT_PASSWORD / MYSQL_RANDOM_ROOT_PASSWORD
MARIADB_ROOT_HOST / MYSQL_ROOT_HOST
MARIADB_MYSQL_LOCALHOST_USER / MARIADB_MYSQL_LOCALHOST_GRANTS
MARIADB_DATABASE / MYSQL_DATABASE
MARIADB_USER / MYSQL_USER, MARIADB_PASSWORD / MYSQL_PASSWORD
MARIADB_INITDB_SKIP_TZINFO / MYSQL_INITDB_SKIP_TZINFO
MARIADB_AUTO_UPGRADE / MARIADB_DISABLE_UPGRADE_BACKUP

提示

有关变量的详细说明请参见 https://hub.docker.com/_/mariadb。

MariaDB 容器的管理与 MySQL 也基本相同，示例命令如下：

```
$ docker run -d --name mariadb1 -v /my/own/datadir:/var/lib/mysql \
  -e MARIADB_ROOT_PASSWORD=my-secret-pw -d mariadb:latest
```

13.4.3 MongoDB

MongoDB 是一种流行的 NoSQL 数据库，使用文档(document)进行数据存储。MongoDB 是无模式(schema-less)的数据库，所以不需要定义数据库模式(schema)，特别适合对写入速度有要求、数据量大的场景。

官方的 Mongo 镜像的 Dockerfile 定义了两个卷：

```
……
VOLUME /data/db /data/configdb

ENV HOME /data/db
COPY docker-entrypoint.sh /usr/local/bin/
ENTRYPOINT ["docker-entrypoint.sh"]

EXPOSE 27017
CMD ["mongod"]
```

下面，做一个简单的实验，使用绑定挂载分配给容器两个卷，示例命令如下：

```
1 $ mkdir -p ~/data/db

2 $ mkdir -p ~/data/configdb
```

```
3 $ docker run -d --name nosql1 -v ~/data/db/:/data/db/ \
    -v ~/data/configdb/:/data/configdb/    mongo
```

在宿主机上创建两个目录，用于保存 MongoDB 的持久数据。

下面进入容器，使用 MongoDB 的客户端软件做一个简单的测试，示例命令如下：

```
$ docker exec -it nosql1 sh

# mongosh
```

显示数据库：

```
test> show databases
admin      8.00 KiB
config     12.00 KiB
local      8.00 KiB
```

创建名为 dbTest 的数据库。

```
test> use dbtest
switched to db dbtest
```

创建名为 myFirstCollection 的集合。

```
dbtest> db.createCollection('myFirstCollection')
{ ok: 1 }
```

在 myFirstCollection 集合中添加文档。

```
dbtest> db.myFirstCollection.insert({item: "card1", qty: 10})
DeprecationWarning: Collection.insert() is deprecated. Use insertOne, insertMany, or bulkWrite.
{
  acknowledged: true,
  insertedIds: { '0': ObjectId("63a81347d17e7efb6321a683") }
}
```

在 myFirstCollection 集合中添加第二个文档。

```
dbtest> db.myFirstCollection.insert({item: "card2", qty: 20})
{
  acknowledged: true,
  insertedIds: { '0': ObjectId("63a8134ed17e7efb6321a684") }
}
```

选择 myFirstCollection 集合中的文档，使用_id 属性显示两个创建的文档。

```
dbtest> db.myFirstCollection.find()
[
  { _id: ObjectId("63a81347d17e7efb6321a683"), item: 'card1', qty: 10 },
  { _id: ObjectId("63a8134ed17e7efb6321a684"), item: 'card2', qty: 20 }
]
```

查询 item 属性为 card1 的文档。

```
dbtest> db.myFirstCollection.find({ item: "card1" })
[
  { _id: ObjectId("63a81347d17e7efb6321a683"), item: 'card1', qty: 10 }
```

```
]
```

查询所有数量属性大于 4 的文档，在本示例中，返回所有文档。

```
dbtest> db.myFirstCollection.find({ qty: { $gt: 4 } })
[
  { _id: ObjectId("63a81347d17e7efb6321a683"), item: 'card1', qty: 10 },
  { _id: ObjectId("63a8134ed17e7efb6321a684"), item: 'card2', qty: 20 }
]
```

查询所有数量属性大于 14 的文档，在本示例中，只有 card2。

```
dbtest> db.myFirstCollection.find({ qty: { $gt: 14 } })
[
  { _id: ObjectId("63a8134ed17e7efb6321a684"), item: 'card2', qty: 20 }
]
```

13.5 习题

请构建一个采用 Apache+MySQL+PHP 架构的容器环境。

参考文献

[1] 任永杰，单海涛. KVM 虚拟化技术：实战与原理解析[M]. 北京：机械工业出版社，2013.

[2] 肖力，汪爱伟，杨俊俊. 深度实践 KVM：核心技术管理运维性能优化与项目实施[M]. 北京：机械工业出版社，2015.

[3] 陈涛. 虚拟化 KVM 极速入门[M]. 北京：清华大学出版社，2022.

[4] 陈涛. 虚拟化 KVM 进阶实践[M]. 北京：清华大学出版社，2022.

[5] 杨保华，戴王剑，曹亚仑. Docker 技术入门与实战[M]. 北京：机械工业出版社，2018.

[6] 奈吉尔 • 波尔顿. 深入浅出 Docker[M]. 李瑞丰，刘康译，译. 北京：人民邮电出版社，2019.

[7] https://libvirt.org/

[8] https://docs.docker.com/

[9] https://www.redhat.com/en/technologies/virtualization/enterprise-virtualization

[10] https://ubuntu.com/server/docs/virtualization-introduction